T0233737

SpringerBriefs in Electrical and Computer Engineering

More information about this series at http://www.springer.com/series/10059

Heng Qi • Keqiu Li

Software Defined Networking Applications in Distributed Datacenters

 Springer

Heng Qi
Dalian University of Technology
Dalian, China

Keqiu Li
Dalian University of Technology
Dalian, China

Microsoft, Encarta, MSN, and Windows are either registered trademarks or trademarks of Microsoft Corporation in the United States and/or other countries.

ISSN 2191-8112   ISSN 2191-8120   (electronic)
SpringerBriefs in Electrical and Computer Engineering
ISBN 978-3-319-33134-8   ISBN 978-3-319-33135-5   (eBook)
DOI 10.1007/978-3-319-33135-5

Library of Congress Control Number: 2016939412

Printed on acid-free paper

This Springer imprint is published by Springer Nature
The registered company is Springer International Publishing AG Switzerland

Preface

Software-defined networking (SDN) has drawn increasing attention from both academia and industry as an emerging network architecture. Compared with closed traditional network architecture, SDN decouples the control function from the forwarding function to build a novel network architecture consisting of three planes: the data plane, control plane, and SDN application. SDN improves the programmability of a network to promote network innovation; however, the basic theories and key technologies of SDN are limited by the initial stage of SDN development. The goal of this book is to provide valuable insights into SDN technologies in distributed datacenters. In particular, we consider three key problems: SDN application design, SDN network deployment, and SDN network management. This book is suitable for SDN researchers and engineers.

In Chap. 1, we introduce the development of SDN and future networks and specifically focus on recent advances in SDN. In Chap. 2, an SDN-based request allocation mechanism is proposed as a typical application of SDN in distributed datacenters. With global information and central control provided by SDN, we propose a joint optimization model for request allocation from the view of both service providers and end-users. Then, we present a Nash bargaining solution (NBS)-based algorithm to implement the request allocation mechanism. In Chap. 3, an SDN controller placement strategy is proposed to achieve the deployment of SDN in distributed datacenters. We formulate the optimal controller placement problem as an integer linear program (ILP) and use an effective approximation algorithm to solve it. In Chap. 4, a management system of heterogeneous SDN controllers is presented to manage the distributed datacenter network. This system shields the differences among heterogeneous controllers to provide a uniform graphical user interface in order to reduce the complexities of network management and SDN application development. Finally, we summarize our studies and highlight future research topics related to SDN in Chap. 5.

We would like to express our appreciation to Professor Sherman Shen and the editors at Springer for their help throughout the publication preparation process. We would also like to thank all of our collaborators for their contributions in this book; in particular, we would like to thank Wenxin Li, Haisheng Yu, Jun Lu, and

Dr. Song Guo. This work was supported by the State Key Program of National Natural Science of China (grant no. 61432002), the National Science Foundation for Distinguished Young Scholars of China (grant no. 61225010), and the Fundamental Research Funds for the Central Universities (grant DUT15QY20).

Dalian, China   Heng Qi
Dalian, China   Keqiu Li
December 2015

Contents

Chapter 1
Introduction

Abstract With the development of computer networks, the defects of traditional Transmission Control Protocol/Internet Protocol (TCP/IP)-based architecture have been amplified. Traditional computer networks are facing big challenges. To break a closed traditional network for eliminating defects and promoting network innovation, software-defined networking (SDN) has been proposed. In this chapter, we illustrate SDN and discuss future network research. We also give a brief overview of recent advances in SDN. Finally, we summarize our work related to SDN in datacenter networks.

1.1 Software-Defined Networking and Future Networks

The scale of computer networks is expanding rapidly, and the demands for cloud, big data, security, and mobility services are always increasing. Traditional network architecture has gradually exposed to defects. To overcome these defects, there is a growing interest in future networks from both academia and industry. Many research projects that propose and design future network architectures have been launched all around the world [1]. The USA has launched a series of future network projects including the NewArch [2], Future Internet Architecture (FIA) [3], Global Environment for Network Innovations (GENI) [4], and Future Internet Design (FIND) [5] projects. In Japan, the New Generation Network (NWGN) project has been implemented, which consists of a series of sub-projects from academia and industry [6]. In the European Union, many future Internet research projects have been launched including the Future Internet Research and Experimentation (FIRE) [7] and Architecture and Design for the Future Internet (4WARD) [8] projects. In China, the China Next Generation Internet (CNGI) project has been established for future network research [9].

From these future network research projects, more and more users have come to the realization that the development of future networks is seriously hindered by the closeness of traditional networks. In general, closed networks face the following problems:

H. Qi, K. Li, *Software-Defined Networking Applications in Distributed Datacenters*, SpringerBriefs in Electrical and Computer Engineering, DOI 10.1007/978-3-319-33135-5_1

- **It is difficult to manage large-scale networks.** Existing networks consist of many closed network devices and a lot of complex networking protocols. There is no one unified, public management platform; hence, network configuration and management must be implemented by specialized network administrators. Moreover, the closeness of network devices increases network management difficulties.
- **It is difficult to guarantee network services.** Traditional networks work in a "best-effort delivery" way. Routers transmit data depending on the current traffic load without a guaranteed quality of service. Because routers are distributed, each router controls data transmission individually. A global network view for traffic scheduling does not exist to improve the quality of service.
- **It is difficult to control network devices.** Existing network devices are black boxes consisting of hardware and software. To preserve their profits, network device vendors are averse to providing standard open interfaces for device control, which reduces flexibility of network control.
- **It is difficult to deploy new network protocols.** Existing closed network devices are designed based on traditional network protocols. It is difficult to implement new protocols without appropriate open application programming interfaces (APIs). A lack of open APIs limits the ability of network programming. Users cannot customize networks based on practical needs, which hinders network innovation.

To address the above problems, closed network architecture should be discouraged and open network architecture should be promoted. To this end, software-defined networking (SDN) has been proposed as a basis for future network construction [10]. SDN is an emerging network architecture that decouples the control function from the forwarding function to break closed network architecture. The framework of SDN enables centralized traffic control and network programmability. Thus, SDN changes the way a network is designed and managed. Figure 1.1 shows the framework of SDN and illustrates its characteristics.

As shown in Fig. 1.1, the framework of SDN usually consists of three planes and two types of APIs. The data plane consists of flow forwarding devices such as switches and routers. It only forwards traffic flows based on the decisions made by the control plane. The control plane consists of controllers (servers), which decide the strategies of traffic flow forwarding. The application plane includes many network applications such as a fire wall and access control list (ACL). To connect these three planes, two types of APIs are used. The first is the open southbound API, and the second is the open northbound API. The open southbound API is the communication interface between the data plane and the control plane. Using the southbound API, the controller can send flow forwarding decisions to the switches. The open northbound API is the network programming interface provided by the control plane. Using the northbound API, the controller can design and implement new network protocols and network innovation applications.

OpenFlow is an important protocol in SDN [11] that facilitates flow table programming in different switches. When the switch vendors integrate the OpenFlow

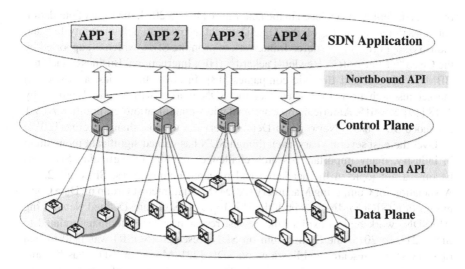

Fig. 1.1 Framework of software-defined networking (SDN)

protocol into their switches, a controller can add and remove flow entries in switches using a standardized interface. OpenFlow enables controllers to realize new traffic policies and schedules while disregarding heterogeneous switches. Since the development of OpenFlow, many users have mistakenly believed that OpenFlow is synonymous with SDN; to be clear, OpenFlow is a successful example of SDN or a typical southbound API in SDN.

1.2 Recent Advances in Software-Defined Networking

The idea of SDN was derived from the Stanford Clean Slate project, namely, Ethane [12]. The goal of Ethane is to build a new architecture for enterprise networks to provide a powerful and simple management model and strong security guarantees [13]. In this project, Martin Casado proposed building one centralized controller by which network administrators can tailor flow-based security policies to their individual requirements. Inspired by this idea, McKeown et al. proposed the concept of OpenFlow in 2008 to enable a network to be programmable [14]. Then, based on OpenFlow, McKeown further proposed SDN in 2009 [15].

As an emerging technology, SDN provides many benefits to overcome the drawbacks of traditional network architectures, which has garnered significant attention from industry. In 2011, the user-driven organization called the Open Networking Foundation (ONF) was founded, which is dedicated to the promotion and adoption of SDN through open standards development [16]. In 2012, the Internet Research Task Force (IRTF) SDN research group (SDNRG) was established to investigate SDN from various perspectives with the goal of identifying the approaches that can be defined, deployed, and used in the near future [17]. In 2013, the Linux Foundation announced the founding of the OpenDaylight Project as a community-led and industry-supported open source framework to promote SDN, which is supported by the Cisco®, Microsoft®, Hewlett-Packard® (HP), International Business Machines (IBM), VMware, and Brocade companies [18]. In 2014, Infonetics, as a telecom market research firm, forecasted that the SDN market will reach 11 billion by 2018 [19]. In 2015, American multinational telecommunications corporation AT&T expanded SDN-based Network-on-Demand services to more than 100 cities [20].

Over the past several years, even though SDN has gained significant momentum in industry, many important research challenges remain. In particular, SDN has become a hot topic in the academia and research communities. Since 2012, the Association for Computing Machinery (ACM) Special Interest Group on Data Communication (SIGCOMM) has built the Hot Topics in Software-Defined Networking (HotSDN) workshop to explore the newest research and developments related to SDN [21]. In 2015, the Symposium on SDN Research (SOSR) was held in lieu of HotSDN for research publications on SDN [22]. Moreover, SDN has become an important topic of interest in many famous academic conferences, such as SIGCOMM, International Conference on Computer Communications (INFOCOM), Networked Systems Design and Implementation (NSDI), and CoNext. Existing work on SDN mainly centers around the three planes of the SDN architecture. Thus, we review state-of-the-art work related to these planes.

1.2.1 Data Plane

The data plane provides rich SDN programming models and abstractions to manage hardware resources [23]. OpenFlow is usually viewed as one general abstraction of the data plane that provides standard interfaces for installing and deleting rules in the flow table [24]. However, there are several drawbacks of OpenFlow. To overcome these drawbacks, other abstractions of the data plane have been proposed.

As a modification of OpenFlow, DevoFlow was proposed to reduce unnecessary costs in order to meet the needs of high-performance networks [25]. Because the abstraction of computation and storage resources are ignored in OpenFlow, Labelcast was proposed to better support future networks by abstracting forwarding resources as well as computation and storage resources in the data plane [26]. The Internet engineering task force (IETF) working group proposed the Forwarding and Control Element Separation (ForCES) protocol to achieve the same objectives as OpenFlow. Compared with OpenFlow, ForCES has a very dynamic model that makes its protocol quite powerful; however, it lacks open source availability for experimentation [27, 28]. Because OpenFlow-based devices are expensive, it is impossible for companies to replace their existing network devices with OpenFlow-based devices. Thus, ClosedFlow was proposed to incorporate techniques of network control over existing devices, while taking advantage of SDN's benefits with no new investment [29]. In recent years, other candidates for the southbound API have been proposed, such as the Path Computation Element (PCE) [30] and the Locator/ID Separation Protocol (LISP) [31].

In addition to abstractions, the flow table is another hot research topic in the data plane. The flow table is the kernel of the OpenFlow switch, which consists of flow entries corresponding to actions. The controller schedules network traffic by installing or deleting flow entries. Because the flow table is implemented with Ternary Content Addressable Memory (TCAM), which is very expensive, the size of the flow table should be as minimal as possible without decreasing the forwarding performance. To address this problem, many solutions have been proposed.

Jiang et al. proposed building a decision forest model for storing more flow entries in less TCAM storage [32]. Kannan et al. proposed Compact TCAM, which reduces the size of the flow entries, thereby managing the large flow table without adding extra TCAM [33]. Soliman et al. proposed source routing techniques to significantly reduce the number of flow table entries [34]. Rifai et al. proposed a compression technique called MINNIE to drastically reduce the number of flow entries with a limited impact on the packet loss rate [35]. Giroire et al. proposed optimizing the size of the forwarding rule space using an integer linear program (ILP) [36]. Huang et al. proposed a partition and allocation algorithm to distribute forwarding rules across all switches with limited TCAM [37]. Zhang et al. proposed the Adaptive Hard Timeout Method (AHTM) to improve the utilization of the flow table by optimizing flow entries' timeouts [38].

1.2.2 Control Plane

The control plane is usually the SDN controller, which is also viewed as the network operating system (OS). In the SDN architecture, the controller connects all forwarding devices in the data plane, by which the network management changes from distribution to centralization. Therefore, a user can tailor network traffic to their particular needs by programming the network using the controller. Moreover, the controller behaves likes a middleware as low-level hardware devices are abstracted. Users can deploy their services on the controller while disregarding low-level devices.

Currently, there are many controllers that include commercial products and open source products. Cisco, VMware, Nicira, Juniper, NEC, and Big Switch Network have their commercial controllers, which usually support their proprietary protocols as well as OpenFlow. The Application Policy Infrastructure Controller (APIC) from Cisco [39] and Big Network Controller from Big Switch Network [40] are two examples. Open source controllers are popular in the academia and research communities, which are usually based on OpenFlow. Examples include NOX, which is implemented in C++, and Floodlight, which is implemented in JAVA.

Released in 2008, NOX was the first OpenFlow controller [41]. The NOX-based network consists of OpenFlow switches, a server running NOX, and a database storing the network view. Since NOX, other OpenFlow controllers have been released. Maestro was proposed to improve the parallelism of the controller [42]. It was the first OpenFlow controller to achieve near linear performance scalability on multi-core processors. Floodlight, a JAVA-based OpenFlow controller, is the core of a commercial controller product from the Big Switch Networks company [43]. It has a simple structure and dexterous operation for SDN beginners. Beacon was created to improve the performance of the controller [44]. Surprisingly, it has high performance and is able to scale linearly with processing cores. McNettle is an extensible controller, which shields the complexity of multi-core processing to preserve a simple programming model [45]. Trema is a full-stack, easy-to-use OpenFlow controller implemented in the Ruby and C languages [46]. RYU is a component-based controller, which provides well-defined APIs for creating applications [47].

The above controllers are centralized controllers, which may cause a single point of failure. To address this problem, the control plane has been designed to be logically centralized but physically distributed. Onix is a distributed control platform, which provides a global view of the network and a general API for control plane implementations [48]. HyperFlow is a distributed event-based control plane, which enables network operators to deploy any number of controllers for tuning the performance of the control plane [49]. DISCO is a distributed multi-domain SDN control plane, which differs from other distributed control planes in that it is adaptable to a heterogeneous, constrained network deployment [50]. The Open Network Operating System (ONOS) is an experimental distributed SDN control platform, which meets the performance and reliability requirements of large-scale networks [51].

Except for research on new controllers, there are few works that compare and evaluate SDN controllers. Shah et al. provided a detailed architectural evaluation of prominent OpenFlow controllers [52]. Based on their evaluations, some promising architectural guidelines have been proposed to improve the scalability of controllers. Fernandez proposed a methodology to compare OpenFlow controllers in reactive or proactive operation paradigms [53]. Monaco et al. proposed incorporating operating system principles into the SDN controller design [54]. Xie et al. presented a detailed survey of existing SDN controllers and analyzed their performance, scalability, and security [55].

1.2.3 Software-Defined Networking Applications

Because SDN enables networks to be programmable, users can build SDN applications to overcome the problems of network management, quality of service (QoS), and routing design. Moreover, users can build SDN applications for datacenter networks, cloud-based networks, mobile networks, and big data.

Network Management. As OpenFlow network evaluation tools, ENVI and SAGE were developed to realize management functionality [56]. Devlic et al. applied SDN in the carrier-grade network, extending the SDN architecture to support multi-provider network management functions in carrier networks [57]. Kim et al. presented an event-driven network control framework, namely, Lithium, which makes network management easier [58]. Sundaresan et al. built a test bed named BISmark for deploying measurements and applications in broadband access networks. This test bed can gather information about network topology, availability, reachability, and so on [59]. Kim et al. designed and implemented Procera to simplify various aspects of network operations and management and serves as the glue between high-level network policies and low-level network configurations [60].

Quality of Service Applications. Kim et al. built a QoS API based on SDN controllers, providing fine-grained automated QoS control in networks [61]. Egilmez et al. proposed the OpenQoS framework to guarantee seamless video delivery with end-to-end QoS support [62, 63]. Moreover, they also designed an optimization framework for QoS routing [64]. Penno et al. proposed the application enabled SDN (A-SDN) framework, which provides QoS and other network services by deploying application-aware network elements [65]. Ko et al. proposed OpenQFlow to support scalable and stateful SDN, providing micro-flow-based QoS [66]. Civanlar et al. described an architecture that can incorporate QoS flows in the OpenFlow environment to support scalable video streaming [67].

Datacenter Networks with Software-Defined Networking. Recent SDN is usually deployed in datacenter networks (DCNs) to improve networking performance. Google deploys SDN in a private wide area network (WAN) that connects datacenters to build the B4 architecture, achieving full control over the whole network to maximize the utilization of links [68]. Tavakoli et al. applied the NOX controller to DCNs, addressing a variety of datacenter requirements [69]. Thanh et al. built a test platform with OpenFlow to measure and analyze energy-aware DCNs [70]. Macapuna et al. proposed a novel datacenter architecture with switching with in-packet bloom filters (SiBF), transforming DCNs into a software problem [71]. Fang et al. used SDN to design a solution for handling datacenter congestion in DCNs [72].

Cloud Computing with Software-Defined Networking. Cloud computing provides Infrastructure-as-a-Service (IaaS), Platform-as-a-Service (PaaS), and Software-as-a-Service (SaaS). However, it is hard to provide Network-as-a-Service (NaaS) because of limited control over network resources. SDN provides powerful user interfaces for controlling networks, which makes it possible to implement NaaS. Feng et al. built a prototype of a cloud-based network by extending the OpenFlow architecture to realize a pay-as-you-go model of network capacity [73]. Benson et al. leveraged SDN techniques to build a novel cloud networking system named CloudNaaS, which provides many virtual networking functions [74]. Banikazemi et al. proposed a novel SDN platform named Meridian to support a network service model for cloud computing [75]. Raghavendra et al. designed a graph algorithm library, which can be loaded into an SDN controller for cloud network management [76].

Other Software-Defined Networking Applications. Caraguay et al. applied SDN to the development of Internet of Things (IoT) applications [77]. Bifulco et al. built a mobile cloud management system based on SDN, which benefits from the dynamic configuration of OpenFlow switches [78]. Wang et al. addressed the problems of mobility in Internet Protocol (IP) networks using SDN while designing an OpenFlow-based mobility protocol [79]. Das et al. presented an SDN-based network management framework named FlowComb for big data processing [80]. Wang et al. integrated a network control function provided by an SDN controller into Hadoop to jointly optimize performance of big data processing and network utilization [81].

1.3   Aim of This Book

This book aims to elaborate on our studies on SDN application design and deployment in distributed datacenters. The remainder of this book is organized as follows.

In Chap. 2, we take advantage of the central control provided by SDN to address the request allocation problem in distributed datacenters. First, we propose a joint optimization model from the view of both service providers and end-users. Second, we present a Nash bargaining solution (NBS)-based method to model the requirements of both the providers and end-users. Third, we develop an efficient algorithm blending the advantages of the logarithmic smoothing technique and the auxiliary variable method. Finally, we conduct many experiments based on real-world traces to show the efficiency of our request allocation algorithm.

In Chap. 3, to deploy SDN in distributed datacenters, we present an effective solution for SDN controller placement. First, we propose a novel placement metric for deploying multiple controllers in large-scale networks. Second, we formulate the optimal controller placement problem as an ILP and use an effective approximation algorithm to find its solution. Finally, we conduct intensive experiments based on many real topologies to demonstrate that our strategy can significantly improve performance over existing methods.

In Chap. 4, to guarantee the performance of network management in distributed datacenters, we design and implement an SDN controller management system. This system consists of the heterogeneous controller management (HCM) module, domain relationships management (DRM) module, database module, and front-end module. It shields the differences between the heterogeneous controllers and provides a unified graphical user interface (GUI) for users and administrators.

In Chap. 5, we summarize our studies regarding SDN application design and deployment in distributed datacenters. We also highlight future research topics related to SDN with the hope of providing valuable insights for researchers and engineers.

References

1. J. Pan, S. Paul and R. Jain. A Survey of the Research on Future Internet Architectures. IEEE Communications Magazine, 2011, 49(7): 26–36.
2. NewArch Project: Future-Generation Internet Architecture, http://www.isi.edu/newarch/.
3. NSF Future Internet Architecture Project, http://www.nets-fia.net/.
4. Global Environment for Network Innovations (GENI) Project, http://www.geni.net/.
5. NSF NETS FIND Project, http://www.nets-find.net.
6. New-Generation Network R&D Project, http://www.nict.go.jp/en/nrh/index.html.
7. FIRE: Future Internet Research and Experimentation, http://cordis.europa.eu/fp7/ict/fire/.
8. The FP7 4WARD Project, http://www.4ward-project.eu/.
9. China Next Generation Internet (CNGI) Project, http://www.cngi.cn/.
10. N. Feamster, J. Rexford and E. Zegura. The Road to SDN: An Intellectual History of Programmable Networks. ACM SIGCOMM Computer Communication Review, 2014, 44(2): 87–98.
11. A. Lara, A. Kolasani and B. Ramamurthy. Network Innovation Using Openflow: A Survey. IEEE Communications Surveys & Tutorials, 2014, 16(1): 493–512.
12. Ethane: A Security Management Architecture, A Stanford Clean Slate Project, http://yuba.stanford.edu/ethane/.
13. M. Casado, M. J. Freedman, J. Pettit, et al. Ethane: Taking Control of the Enterprise. ACM SIGCOMM Computer Communication Review, 2007, 37(4): 1–12.
14. N. McKeown, T. Anderson, H. Balakrishnan, et al. OpenFlow: Enabling Innovation in Campus Networks. ACM SIGCOMM Computer Communication Review, 2008, 38(2): 69–74.
15. N. McKeown. Software-defined Networking. INFOCOM Keynote Talk, Rio de Janeiro, Brazil, April 2009.
16. Open Networking Foundation (ONF), https://www.opennetworking.org/about/onf-overview.
17. Software-Defined Networking Research Group (SDNRG), https://irtf.org/sdnrg.
18. L. Foundation, Opendaylight: An Open Source Community and Meritocracy for Software-Defined Networking, A Linux Foundation Collaborative Project, April 2013. http://www.opendaylight.org/resources/publications.
19. Infonetics, Carrier SDN and NFV Hardware and Software Market Size and Forecast Report, November 2014. http://www.infonetics.com/pr/2014/Carrier-SDN-NFV-Market-Highlights.asp.
20. AT & T, A Software-Centric Network – Network on Demand & Universal CPE. http://about.att.com/innovation/showcase/networkondemand.
21. ACM SIGCOMM Workshop on Hot Topics in Software Defined Networking (HotSDN). http://conferences.sigcomm.org/sigcomm/2014/hotsdn.php.
22. ACM SIGCOMM Symposium ON SDN Research (SOSR). http://www.sigcomm.org/events/SOSR.
23. M. Casado, N. Foster and A. Guha. Abstractions for Software-Defined Networks. Communications of the ACM, 2014, 57(10): 86–95.
24. L. Schiff, M. Borokhovich and S. Schmid. Reclaiming the Brain: Useful OpenFlow Functions in the Data Plane. Proceedings of the 13th Workshop on Hot Topics in Networks, ACM, 2014: 1–7.
25. A. R. Curtis, J. C. Mogul, J. Tourrilhes, et al. DevoFlow: Scaling Flow Management for High-performance Networks. ACM SIGCOMM Computer Communication Review, 2011, 41(4): 254–265.
26. J. Su, G. Lv, Z. Sun, et al. Labelcast: A Novel Data Plane Abstraction for SDN. Open Networking Summit, Santa Clara, USA, April 2013.
27. A. Doria, J. H. Salim, R. Haas, et al. Forwarding and Control Element Separation (ForCES) Protocol Specification. [Online]. Available: http://tools.ietf.org/html/rfc5810.
28. D. Kreutz, F. M. V. Ramos, P. E. Verissimo, et al. Software-Defined Networking: A Comprehensive Survey. Proceedings of the IEEE, 2015, 103(1): 14–76.

29. R. Hand and E. Keller. ClosedFlow: OpenFlow-Like Control over Proprietary Devices, Proceedings of the 3rd Workshop on Hot Topics in Software Defined Networking. ACM, 2014: 7–12.
30. A. Farrel, J. P. Vasseur and J. Ash. A Path Computation Element (PCE)-Based Architecture. [Online]. Available: http://tools.ietf.org/html/rfc4655.
31. D. Farinacci, V. Fuller, D. Meyer and D. Lewis. The Locator/ID Separation Protocol (LISP). [Online]. Available: https://tools.ietf.org/html/rfc6830.
32. W. Jiang, V. K. Prasanna and N. Yamagaki. Decision Forest: A Scalable Architecture for Flexible Flow Matching on FPGA. Proceedings of the 2010 International Conference on Field Programmable Logic and Applications (FPL), IEEE, 2010: 394–399.
33. K. Kannan and S. Banerjee. Compact TCAM: Flow Entry Compaction in TCAM for Power Aware SDN. Distributed Computing and Networking, Springer Berlin Heidelberg, 2013: 439–444.
34. M. Soliman, B. Nandy, I. Lambadaris and P. Ashwood-Smith. Source Routed Forwarding with Software Defined Control, Considerations and Implications. Proceedings of the 2012 ACM Conference on CoNEXT Student Workshop. ACM, 2012: 43–44.
35. M. Rifai, N. Huin, C. Caillouet, et al. Too Many SDN Rules? Compress Them with MINNIE. Proceedings of the 2015 Global Communications Conference (GLOBECOM), IEEE, 2015.
36. F. Giroire, J. Moulierac and T. K. Phan. Optimizing Rule Placement in Software-Defined Networks for Energy-Aware Routing. Proceedings of the 2014 Global Communications Conference (GLOBECOM), IEEE, 2014: 2523–2529.
37. J. Huang, G. Chang, C. Wang and C. Lin. Heterogeneous Flow Table Distribution in Software-defined Networks. IEEE Transactions on Emerging Topics in Computing, DOI: 10.1109/TETC.2015.2457333.
38. L. Zhang, R. Lin, S. Xu and S. Wang. AHTM: Achieving Efficient Flow Table Utilization in Software Defined Networks. Proceedings of the 2014 Global Communications Conference (GLOBECOM), IEEE, 2014: 1897–1902.
39. Cisco Application Policy Infrastructure Controller (APIC). http://www.cisco.com/c/en/us/products/cloud-systems-management/application-policy-infrastructure-controller-enterprise-module/index.html.
40. Big Network Controller. http://bigswitch.com/products/SDN-Controller.
41. N. Gude, T. Koponen, J. Pettit, et al. NOX: Towards An Operating System for Networks. ACM SIGCOMM Computer Communication Review, 2008, 38(3): 105–110.
42. Z. Cai, A. L. Cox and T. S. Eugene Ng. Maestro: A System for Scalable Openflow Control. Rice University Technical Report TR11-07, December 2011.
43. Floodlight: An Open SDN Controller. http://www.projectfloodlight.org/floodlight/.
44. D. Erickson. The Beacon Openflow Controller. Proceedings of the 2nd ACM SIGCOMM Workshop on Hot Topics in Software Defined Networking. ACM, 2013: 13–18.
45. A. Voellmy and J. Wang. Scalable Software Defined Network Controllers, ACM SIGCOMM Computer Communication Review, 2012, 42(4): 289–290.
46. Trema: Openflow Controller. https://trema.github.io/trema/.
47. Ryu: A Component-based Software Defined Networking Framework. http://osrg.github.io/ryu/.
48. T. Koponen, M. Casado, N. Gude, et al. Onix: A Distributed Control Platform for Large-scale Production Networks. Proceedings of Operating Systems Design and Implementation (OSDI). USENIX Association, 2010.
49. A. Tootoonchian and Y. Ganjali. Hyperflow: A Distributed Control Plane for Openflow. Proceedings of the 2010 Internet Network Management Conference on Research on Enterprise Networking. USENIX Association, 2010: 3–3.
50. K. Phemius, M. Bouet and J. Leguay. Disco: Distributed Multi-domain SDN Controllers. Proceedings of Network Operations and Management Symposium (NOMS). IEEE/IFIP, 2014: 1–4.
51. P. Berde, M. Gerola, J. Hart, et al. ONOS: Towards An Open, Distributed SDN OS. Proceedings of the 3rd Workshop on Hot Topics in Software Defined Networking. ACM, 2014: 1–6.

52. S. A. Shah, J. Faiz, M. Farooq, et al. An Architectural Evaluation of SDN Controllers. the 2013 International Conference on Communications (ICC). IEEE, 2013: 3504–3508.

53. M. P. Fernandez. Comparing Openflow Controller Paradigms Scalability: Reactive and Proactive. The 27th International Conference on Advanced Information Networking and Applications (AINA). IEEE, 2013: 1009–1016.

54. M. Monaco, O. Michel and E. Keller. Applying Operating System Principles to SDN Controller Design. Proceedings of the 12th Workshop on Hot Topics in Networks. ACM, 2013.

55. J. Xie, D. Guo, Z. Hu, et al. Control Plane of Software Defined Networks: A Survey. Computer Communications, 2015, 67: 1–10.

56. A. Hakiri, A. Gokhale, P. Berthou, D. C. Schmidt and T. Gayraud. Software-Defined Networking: Challenges and Research Opportunities for Future Internet. Computer Networks, 2014, 75: 453–471.

57. A. Devlic, W. John and P. Skoldstrom. Carrier-grade Network Management Extensions to the SDN Framework. Proceedings of the 8th Swedish National Computer Networking Workshop (SNCNW), Stockholm, Sweden. 2012.

58. H. Kim, A. Voellmy, S. Burnett, N. Feamster and R. Clark. Lithium: Event-Driven Network Control. Georgia Institute of Technology Technical Report, 2012.

59. S. Sundaresan, S. Burnett, N. Feamster and W. Donato. BISmark: A Testbed for Deploying Measurements and Applications in Broadband Access Networks. Proceedings of the USENIX Annual Technical Conference (USENIX ATC 14). USENIX, 2014: 383–394.

60. H. Kim and N. Feamster. Improving Network Management with Software Defined Networking. IEEE Communications Magazine, 2013, 51(2): 114–119.

61. W. Kim, P. Sharma, J. Lee, et al. Automated and Scalable QoS Control for Network Convergence. Proceedings of USENIX INM/WREN 2010, San Jose, CA, April 2010.

62. H. E. Egilmez, S. T. Dane, K. T. Bagci and A. M. Tekalp. OpenQoS: An OpenFlow Controller Design for Multimedia Delivery with End-to-End Quality of Service over Software-Defined Networks. 2012 Asia-Pacific Signal & Information Processing Association Annual Summit and Conference (APSIPA ASC). IEEE, 2012: 1–8.

63. H. E. Egilmez, S. T. Dane, B. Gorkeml and A. M. Tekalp. Openqos: Openflow Controller Design and Test Network for Multimedia Delivery with Quality of Service. Proceedings of NEM Summit, Implementing Future Media Internet Towards New Horiz, 2012: 22–27.

64. H. E. Egilmez, B. Gorkeml, A. M. Tekalp and S. Civanlar. Scalable Video Streaming over OpenFlow Networks: An Optimization Framework for QoS Routing. Proceedings of the 18th International Conference on Image Processing (ICIP). IEEE, 2011: 2241–2244.

65. R. Penno, T. Reddy, M. Boucadair, D. Wing and S. Vinapamula. Application Enabled SDN (A-SDN). [Online]. Available: https://tools.ietf.org/html/draft-penno-pcp-asdn-00.

66. N. S. Ko, H. HEO, J. D. PARK and H. S. PARK. OpenQFlow: Scalable OpenFlow with Flow-Based QoS. IEICE TRANSACTIONS on Communications, 2013, E96-B(2): 479–488.

67. S. Civanlar, M. Parlakisik, A. M. Tekalp, et al. A QoS-Enabled OpenFlow Environment for Scalable Video Streaming. Proceedings of the IEEE GLOBECOM Workshops on Network of the Future. IEEE, 2010: 351–356.

68. S. Jain, A. Kumar, S. Mandal, et al. B4: Experience with A Globally-deployed Software Defined WAN. ACM SIGCOMM Computer Communication Review. ACM, 2013, 43(4): 3–14.

69. A. Tavakoli, M. Casado, T. Koponen and S. Shenker. Applying NOX to the Datacenter. Proceedings of the 8th Workshop on Hot Topics in Networks. ACM, 2009.

70. N. H. Thanh, P. N. Nam, T. H. Truong, et al. Enabling Experiments for Energy-efficient Data Center Networks on OpenFlow-based Platform. Proceedings of the 4th International Conference on Communications and Electronics (ICCE). IEEE, 2012: 239–244.

71. C. Macapuna, C. E. Rothenberg and M. F. Magalhaes. In-Packet Bloom Filter based Data Center Networking with Distributed OpenFlow Controllers. Proceedings of the IEEE GLOBECOM Workshops on Management of Emerging Networks and Services. IEEE, 2010: 584–588.

72. S. Fang, Y. Yu, C. H. Foh, et al. A Loss-Free Multipathing Solution for Data Center Network using Software-Defined Networking Approach. Proceedings of the Asia-Pacific Magnetic Recording Conference, Digest APMRC. IEEE, 2012: 1–8.
73. T. Feng, J. Bi, H. Hu and H. Cao. Networking as A Service: A Cloud-based Network Architecture. Journal of Networks, 2011, 6(7): 1084–1090.
74. T. Benson, A. Akella, A. Shaikh and S. Sahu. CloudNaaS: A Cloud Networking Platform for Enterprise Applications. Proceedings of the 2nd ACM Symposium on Cloud Computing. ACM, 2011.
75. M. Banikazemi, D. Olshefski, A. Shaikh, J. Tracey and G. Wang. Meridian: An SDN Platform for Cloud Network Services. IEEE Communications Magazine, 2013, 51(2): 120–127.
76. R. Raghavendra, J. Lobo and K. W. Lee. Dynamic Graph Query Primitives for SDN-based Cloudnetwork Management. Proceedings of the 1st Workshop on Hot topics in Software Defined Networks. ACM, 2012: 97–102.
77. A. Caraguay, A. Peral, L. Lopez and L. Villalba. SDN: Evolution and Opportunities in the Development IoT Applications. International Journal of Distributed Sensor Networks, http://dx.doi.org/10.1155/2014/735142, 2014.
78. R. Bifulco, M. Brunner, R. Canonico, et al. Scalability of A Mobile Cloud Management System. Proceedings of the 1st MCC Workshop on Mobile Cloud Computing. ACM, 2012: 17–22.
79. Y. Wang and J. Bi. A Solution for IP Mobility Support in Software Defined Networks. Proceedings of the 23rd International Conference on Computer Communication and Networks (ICCCN). IEEE, 2014: 1–8.
80. A. Das, C. Lumezanu, Y. Zhang, et al. Transparent and Flexible Network Management for Big Data Processing in the Cloud. Proceedings of the 5th USENIX Workshop on Hot Topics in Cloud Computing (HotCloud'13). USENIX, 2013: 1–6.
81. G. Wang, T. S. Ng, A. Shaikh. Programming Your Network at Run-time for Big Data Applications. Proceedings of the 1st Workshop on Hot topics in Software Defined Networks. ACM, 2012: 103–108.

Chapter 2
Software-Defined Networking Based Request Allocation in Distributed Datacenters

Abstract Large-scale Internet applications, such as information retrieval or video streaming, are usually built on top of distributed datacenters. In these applications, the *request allocation* problem is a fundamental problem, aiming to efficiently allocate massive requests among distributed datacenters. Generally, there are two basic factors that should be considered. First, from an overall system perspective, service provider expects to achieve *high bandwidth utilization* and *load balance*. Second, from an individual perspective, end-users have a strong desire for *good user experience* and *fair treatment*. To the best of our knowledge, existing approaches solely focus on either the former or the latter. Software-defined networking (SDN) makes it possible to implement global optimization over an entire network consisting of distributed datacenters. Thus, an SDN controller can be used as the central portal to allocate requests, satisfying the needs of both service providers and end-users. To address this problem, we first develop a general formulation of the request allocation problem. Specifically, we guarantee the benefits of both the service providers and end-users, which are modeled by two Nash bargaining games. Then, we further present an efficient request allocation algorithm based on logarithmic smoothing. We theoretically prove that our request allocation algorithm significantly converges to a unique solution. Finally, we conduct a large number of experiments based on real-world traces. These simulation results demonstrate the efficiency of our request allocation algorithm.

Large-scale Internet applications, such as video streaming (Netflix), information retrieval (Google), and social networking (Facebook), provide service to hundreds of millions of end-users. To guarantee both reliability and performance of services, providers must deploy distributed datacenters to deal with ten million requests every day. Once applications are deployed on distributed datacenters, the request allocation problem is of great importance to both service providers and end-users.

As shown in Fig. 2.1, a simple model consists of two end-users and two datacenters deployed by a service provider. The request allocation problem becomes how to efficiently distribute massive requests among distributed datacenters through a centralized traffic controller. There are two basic goals that must be considered from the view of both the service provider and end-users. First, high bandwidth

© The Author(s) 2016

H. Qi, K. Li, *Software-Defined Networking Applications*
in Distributed Datacenters, SpringerBriefs in Electrical and Computer Engineering,
DOI 10.1007/978-3-319-33135-5_2

Fig. 2.1 Distributed
datacenters model

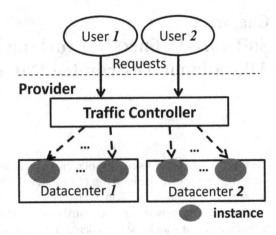

utilization and load balance must be achieved from the overall system perspective of the service provider. Second, end-users have a strong desire for good user experience and fair treatment from their individual perspective.

There are two observations that motivate these goals. First, if large costs have been incurred, the benefit of the service provider must be guaranteed. For example, Google has already spent a lot of money to deploy more than 450,000 servers in its thirty datacenters [1]. Overloading can lead to poor performance and is vulnerable to failures [2]. To be profitable for service providers, high bandwidth utilization and load balance among multiple datacenters must be achieved. Second, as business increases, end-user demands for low response time, fairness, and low costs also increase [3]. Thus, good user experience and fairness among end-users must be ensured. Unfortunately, previous works can be generalized as a continuous optimization problem that mainly concentrates on part of the goals defined in different ways, i.e., minimizing the total cost for service providers or fairness among end-users; such approaches involve several iterations [4, 7].

Existing work on request allocation either solely focuses on service providers or end-users, i.e., minimizing total costs for service providers or fairness among end-users. For example, from a provider's point of view, Qureshi et al. proposed a simple cost-aware request allocation policy that utilizes geographical diversity of electricity prices to preferentially allocate requests to datacenters where energy is cheaper [4]. Gao et al. designed FORTE, a request-routing framework that provides a three-way trade-off between access latency, electricity cost, and carbon footprint [5]. Liu et al. considered the effects of request allocation on providing environment gains by using green energy [6]. Xu et al. developed an efficient request allocation algorithm that considered both bandwidth and electricity costs [7]. Boloor et al. proposed a novel approach of data-oriented dynamic service-request allocation with gi first-in, first-out (gi-FIFO) scheduling to globally increase the profit charged by cloud computing systems [8].

There is also existing work that takes end-user benefits into account. Wendell et al. developed a decentralized request allocation algorithm for cloud services and evaluated its performance using a prototype and realistic traffic traces [9]. Xu et al.

discussed fairness between users in the process of request allocation, and they also developed a request allocation algorithm based on the alternating direction method of multipliers (ADMM), which efficiently balances the trade-off between performance and cost [10].

Unlike existing work, we want to take both the benefit to the service provider and benefit to end-users into account. To give an optimal request allocation solution, the global information of the entire network is needed. Fortunately, software-defined networking (SDN) makes it feasible to collect global network information and control the entire network [11]. Inspired by this, we construct our request allocation model based on the SDN controller. To guarantee both benefits to the service provider and end-users, we apply two Nash bargaining games where each Nash product denotes the benefit of the service provider and end-users, respectively. Furthermore, we propose a concept named user experience degree to qualify user experience that depends on the most important metric response time and some inherent cost. To obtain the final solution, we present an efficient request allocation algorithm based on logarithmic smoothing. We theoretically prove that our logarithmic smoothing-based request allocation algorithm significantly converges to a unique solution [12].

Our main contributions are summarized as follows:

- We present a general formulation of request allocation for a scenario of distributed datacenters and multiple end-users.
- We present a Nash product-based method to capture both benefits of the service provider and end-users.
- We propose an efficient request algorithm based on logarithmic smoothing and theoretically prove that our request allocation algorithm significantly converges to a unique solution.
- We evaluate the efficiency of our request allocation algorithm using real-world traces in our simulation.

The remainder of this chapter is organized as follows. In Sect. 2.1, we show a framework for request allocation. In Sect. 2.2, we give our request allocation algorithm based on logarithmic smoothing. In Sect. 2.3, we discuss the experiment evaluation. Finally, we conclude this chapter in Sect. 2.4.

2.1 A Software-Defined Networking Framework for Request Allocation

2.1.1 Infrastructure

Consider that the service provider deploys a set of datacenters $\mathcal{M}$ for better reliability and performance, $\mathcal{M} = \{d_1, d_2, \ldots, d_M\}$. Each datacenter $d_j \in M$ is equipped with a fixed bandwidth capacity U_j. Let $\mathcal{K} = \{c_1, c_2, \ldots, c_K\}$ denote the set of end-users and $\mathcal{N}$ denote the set of application instances offered by service provider, $\mathcal{N} = \{a_1, a_2, \ldots, a_N\}$. Let b_i be the amount of bandwidth to handle one request when serving $a_i \in \mathcal{N}$. Generally, service providers make several copies for each application instance. The set of application instances hosted by d_j is defined as $w_j = (w_{1,j}, w_{2,j}, \ldots, w_{N,j})$, where $w_{i,j}$ is the binary variable that indicates whether a_i is located on d_j. Then, we have

$$w_{i,j} = \begin{cases} 1 & \text{if } a_i \text{ is located on } d_j, \\ 0 & \text{otherwise.} \end{cases}$$

For all $a_i \in \mathcal{N}$, assume that $c_k \in \mathcal{K}$ can only make one request for a given moment. End-users' requests are described by a matrix $r_N = [r_{k,i}]_{K \times N}$, where $r_{k,i}$ denotes whether c_k requests a_i. Clearly, this implies

$$r_{k,i} = \begin{cases} 1 & \text{if } c_k \text{ requests } a_i, \\ 0 & \text{otherwise.} \end{cases}$$

Let $v_{i,j}^k$ represent whether d_j gets the request from c_k to a_i. In this case, we have

$$v_{i,j}^k = \begin{cases} 1 & \text{if } r_{k,i} w_{i,j} \neq 0 \text{ and } d_j \text{ gets the} \\ & \text{request from } c_k \text{ to } a_i, \\ 0 & \text{otherwise.} \end{cases} \tag{2.1}$$

$v_{i,j}^k$ is our goal such that each request is allocated to a suitable datacenter with both benefits of the service provider and end-users guaranteed. We apply Nash bargaining solutions [14] to interpret the benefits of service providers and end-users. The Nash bargaining solution is known as a non-zero-sum game, where players cooperate to achieve a win–win solution so that social utility gains are maximized and commodities owned by each player do not exceed its capacity. This corresponds to maintaining fairness and high utility for players.

2.1.2 Service Provider and End-Users

We now present methods for defining both benefits of service providers and end-users.

2.1.2.1 The Benefit of Service Providers

The service provider incurs large costs building $\mathcal{M}$ datacenters. Therefore, the service provider expects to achieve *high bandwidth utilization* and *load balance* from an overall system perspective. High bandwidth utilization contributes to improved throughput. One can imagine an overloading case of any one datacenter, where the response time can be significantly increased and the throughput of the overall system can be seriously decreased. To capture bandwidth resource usage at each datacenter, we define the bandwidth utilization of each datacenter d_j as P_j, which is given by

$$P_j = \sum_{k=1}^{K} \sum_{i=1}^{N} \frac{r_{k,i} w_{i,j} v_{i,j}^k b_i}{U_j}.$$

At this time, we interpret the benefit of the service provider as the social utility gain in a Nash bargaining game. The Nash bargaining game is described as follows: M datacenters are viewed as players, and requests from K end-users are viewed as commodities. Each player enters the game with bandwidth utilization P_j as the utility function and aims to maximize its own utility. In the long-term, it achieves a win–win solution. By utilizing the concept of the Nash bargaining solution, we simply obtain an equilibrium point that guarantees the load balance between multiple datacenters and achieves high bandwidth utilization. The benefit of the service provider is computed by $\prod_{j=1}^{M} P_j$.

2.1.2.2 The Benefit of End-Users

From the perspective of end-users, they have a strong desire for *good user experience* and *fair treatment*. User experience is end-users' perceptions and responses of system aspects such as utility, ease of use, and efficiency, that result from the use or anticipated use of service. In our framework, we consider that user experience depends on the most important metric response time inside the datacenter and some inherent cost, i.e., the price of application instance, electricity cost, etc.

We build a set of parallel queues denoting each application instance inside datacenters, such as $queue_{i,j}$. Requests arrive one by one into its corresponding infinite capacity queue. For each application instance in d_j, let $\theta_{i,j}$ denote the mean arrival rate of requests and $\tau_{i,j}$ represent the inter-arrival time whose expected value is $1/\theta_{i,j}$; $\varphi_{i,j}$ captures the mean service time, and $\rho_{i,j} = \theta_{i,j}\varphi_{i,j}$ is the traffic offered corresponding to the fraction of time the server is busy if the application instance is served by a single server. Let $\sigma_{\varphi_{i,j}}^2$ and $\sigma_{\tau_{i,j}}^2$ represent the squared coefficient of variation of service time and request inter-arrival time, respectively. Now, we obtain an expression for the average response time. We approximate $queue_{i,j}$ by a G/G/1 queue using the method in [15]. In this case, the average response time is defined by

$$\delta_{i,j} = \varphi_{i,j} + \varphi_{i,j} \frac{\rho_{i,j}}{1 - \rho_{i,j}} \left(\frac{\sigma_{\varphi_{i,j}}^2 + \sigma_{\tau_{i,j}}^2}{2} \right),$$

where the first term is the average service time and the second term denotes the average waiting time in $queue_{i,j}$.

Next, we present the inherent cost for end-users. Let $\xi_{i,j}^k$ represent the inherent cost of c_k requests a_i within d_j. The inherent cost may be caused by the price of application, spatial distance from end-user to datacenter, or electricity cost. In order to qualify user experience, we propose the concept of user experience degree, which ranges from 0 to 1. Let the user experience degree of c_k when requesting the application instance a_i within d_j be denoted by $u_{i,j}^k$, which is defined as follows:

$$u_{i,j}^k = \exp(-\delta_{i,j} - \xi_{i,j}^k).$$

Note that $u_{i,j}^k$ is a decreasing function of the average response time $\delta_{i,j}$ and inherent cost $\xi_{i,j}^k$, which is reasonable in practice. Based on the average response time $\delta_{i,j}$ and inherent cost $\xi_{i,j}^k$, the average user experience degree of end-user c_k is denoted by T_k, which is given by

$$T_k = \sum_{j=1}^{M} \sum_{i=1}^{N} r_{k,i} w_{i,j} v_{i,j}^k u_{i,j}^k \Big/ \sum_{i=1}^{N} r_{k,i}.$$

We are now in a position to formally formulate the benefit of end-users as the Nash product in another Nash bargaining game. In this game, K end-users are viewed as players and M datacenters are viewed as commodities. Rational players will seek to an appropriate datacenter to maximize its utility, which is denoted by its own average user experience T_k. Players cooperate in this game to achieve a win–win solution. Therefore, fairness among end-users and high user experience can be achieved from the Nash bargaining solution. Thus, the benefit of end-users is defined by $\prod_{k=1}^{K} T_k$.

2.1.3 Problem Formulation

We now formulate the request allocation problem as an optimization problem, considering both the benefits of service provider and end-users.

$$\max \left(\frac{\prod_{j=1}^{M} P_j}{\prod_{k=1}^{K} T_k} \right) \tag{2.2}$$

$$\text{s.t.} \quad \sum_{k=1}^{K} \sum_{i=1}^{N} r_{k,i} w_{i,j} v_{i,j}^k b_i \le U_j, \forall j, \tag{2.3}$$

$$\sum_{j=1}^{M} v_{i,j}^k = 1, \forall k, \forall i, \tag{2.4}$$

where Eq. 2.3 stands for the bandwidth capacity constraint, and Eq. 2.4 denotes that each request can only be allocated to one datacenter. Here, we ignore the case when $r_{k,i} * w_{i,j} = 0$ while $v_{i,j}^k = 1$ since it has no influence on the two objectives. We obtain the real value of $v_{i,j}^k$ from Eq. 2.1. The first objective in Eq. 2.2 stands for the benefit of the service provider, and the second represents the benefit of end-users. By taking the logarithm of the two objectives in Eq. 2.2, we derive another multi-objective optimization problem as follows:

$$
\max \left(
\begin{array}{c}
\sum_{j=1}^{M} \ln \sum_{k=1}^{K} \sum_{i=1}^{N} \frac{r_{k,i} w_{i,j} v_{i,j}^k b_i}{U_j} \\
\sum_{k=1}^{K} \ln \frac{\sum_{j=1}^{M} \sum_{i=1}^{N} r_{k,i} w_{i,j} v_{i,j}^k u_{i,j}^k}{\sum_{i=1}^{N} r_{k,i}}
\end{array}
\right).
\tag{2.5}
$$

By utilizing the linear weighted sum method in Eq. 2.5, we obtain a single objective optimization problem. Note that by adding a weight factor in front of each objective, any desired trade-off point between the benefit of the service provider and the benefit of end-users can be achieved. Let λ_1 and λ_2 denote the weight of the benefit of the service provider and benefit of end-users, respectively. The sum of these two weights should be a unitary value, i.e., $\lambda_1 + \lambda_2 = 1$. Then, obtain the following single objective optimization problem:

$$
\max \quad H(v),
\tag{2.6}
$$

where

$$
H(v) = \lambda_1 \sum_{j=1}^{M} \ln \sum_{k=1}^{K} \sum_{i=1}^{N} \frac{r_{k,i} w_{i,j} v_{i,j}^k b_i}{U_j} +
$$

$$
\lambda_2 \sum_{k=1}^{K} \ln \frac{\sum_{j=1}^{M} \sum_{i=1}^{N} r_{k,i} w_{i,j} v_{i,j}^k u_{i,j}^k}{\sum_{i=1}^{N} r_{k,i}}.
$$

The following theorem indicates that the optimal solution to Eq. 2.6 must be the efficient solution to Eq. 2.5.

Theorem 1. *The optimal solution to the single objective optimization problem in Eq. 2.6 must be the efficient solution to the multi-objective optimization problem in Eq. 2.5.*

Proof. First, let

$$
\Psi_1(v) = \sum_{j=1}^{M} \ln \sum_{k=1}^{K} \sum_{i=1}^{N} \frac{r_{k,i} w_{i,j} v_{i,j}^k b_i}{U_j},
$$

and

$$
\Psi_2(v) = \sum_{k=1}^{K} \ln \frac{\sum_{j=1}^{M} \sum_{i=1}^{N} r_{k,i} w_{i,j} v_{i,j}^k u_{i,j}^k}{\sum_{i=1}^{N} r_{k,i}}.
$$

Let v be the matrix $(v_{i,j}^k)$ and the feasible region of v be denoted by Z, which is a set of v that satisfies the constraint in Eqs. 2.3 and 2.4. We assume that v^* is the optimal solution to the single objective optimization problem in Eq. 2.6, but is not the efficient solution to the multi-objective optimization problem in Eq. 2.5. Then, there exists $v^* \in Z$ such that

$$\Psi_1(v^*) > \Psi_1(v^*),$$
$$\Psi_2(v^*) > \Psi_2(v^*).$$

Then, we obtain

$$\lambda_1 \Psi_1(v^*) + \lambda_2 \Psi_2(v^*) > \lambda_1 \Psi_1(v^*) + \lambda_2 \Psi_2(v^*).$$

This indicates that v^* is not the optimal solution to the problem in Eq. 2.6. Hence, the theorem is proven.

2.2 Request Allocation Algorithm with a Software-Defined Networking Global View

Clearly, the problem in Eq. 2.6, a nonlinear integer optimization problem whose variables can only take on integer quantities or discrete values, is NP-hard [13]. Problems like this are usually solved by implicit enumeration methods or cutting plane methods. However, the computational complexity may increase significantly as the number of discrete variables increases. In this paper, we resort to logarithmic smoothing [16], a continuation approach for nonlinear integer programming problems. We first present logarithmic smoothing and then detail our request allocation algorithm.

2.2.1 Logarithmic Smoothing

Logarithmic smoothing can be an efficient smoothing method. Before presenting the logarithmic smoothing technique, we make a few transformations. First, let X be the matrix $(X_j^{k,i})$, where $X_j^{k,i}$ is a number between 0 and 1 and is defined by

$$X_j^{k,i} = \begin{cases} r_{k,i} w_{i,j} v_{i,j}^k & \text{if } r_{k,i} w_{i,j} \neq 0, \\ 0 & \text{otherwise.} \end{cases} \tag{2.7}$$

Next, we include the slack variable Y_j for Eq. 2.3. Then Eq. 2.3 can be written in matrix form as follows:

$$A_1 X + Y = U.$$

The matrices A_1, X, and Y are defined by

$$A_1 = \begin{bmatrix} B^T & 0 & 0 & 0 \\ 0 & B^T & 0 & 0 \\ \vdots & \vdots & \ddots & \vdots \\ 0 & 0 & 0 & B^T \end{bmatrix},$$

$$X = (\overbrace{\underbrace{X_1^{1,1},\ldots,X_1^{1,N}}_{N},\ldots,X_1^{2,1},\ldots,X_1^{2,N}}^{K},\ldots,X_M^{K,N})^T,$$

$$Y = (Y_1,\ldots,Y_M)^T,$$

$$U = (U_1,\ldots,U_M)^T,$$

where A_1 is an $(M \times MKN)$ matrix and

$$B = (\overbrace{\underbrace{b_1,\ldots,b_N}_{K},\ldots,b_1,\ldots,b_N}^{N})^T.$$

Equation 2.4 can also be transformed into the following matrix form:

$$A_2 X = e.$$

Here, A_2 and e are defined by

$$A_2 = (\overbrace{\underbrace{\hat{e}_1,\ldots,\hat{e}_{KN}}_{KN},\ldots,\hat{e}_1,\ldots,\hat{e}_{KN}}^{M}),$$

$$e = [1,1,\ldots,1]^T,$$

where A_2 is a $(KN \times MKN)$ matrix and $\hat{e}_i$ is the ith column of the $(KN \times KN)$ identity matrix I_{KN}; e is a vector with KN elements.

Therefore, we transform Eq. 2.6 into a matrix form in Eq. 2.8 that contains equality constraints only. Then, we have

$$\min \quad -f(X) \tag{2.8}$$
$$\text{s.t.} \quad A_1 X + Y = U,$$
$$A_2 X = e,$$
$$X \in \{0,1\}^{M \times K \times N},$$

where

$$f(X) = \lambda_1 \sum_{j=1}^{M} \ln \sum_{k=1}^{K} \sum_{i=1}^{N} \frac{X_j^{k,i} b_i}{U_j} +$$

$$\lambda_2 \sum_{k=1}^{K} \ln \frac{\sum_{j=1}^{M} \sum_{i=1}^{N} X_j^{k,i} u_{i,j}^k}{\sum_{i=1}^{N} r_{k,i}}.$$

Now, we present logarithmic smoothing, which primarily relies on a logarithmic barrier and penalty term. Let $\Phi(X,Y)$ be the smoothing function defined by

$$\Phi(X,Y) = -\sum_{j=1}^{M} \sum_{k=1}^{K} \sum_{i=1}^{N} [\ln X_j^{k,i} + \ln(1 - X_j^{k,i})] + \sum_{j=1}^{M} \ln Y_j.$$

Since the feasible region of the independent variable must be nonnegative in the logarithm function, the logarithmic barrier function simply eliminates inequality constraints. Thus, we obtain the transformed problem as follows:

$$\min \quad -f(X) + \mu \Phi(X,Y) \tag{2.9}$$

$$\text{s.t.} \quad A_1 X + Y = U,$$

$$A_2 X = e,$$

where $\mu > 0$ is the smoothing parameter and $-f(X) + \mu \Phi(X,Y)$ is strictly convex if μ is large enough as prescribed by Lemma 3.1 in [17]. The following theorem indicates that there is a unique solution to Eq. 2.9.

Theorem 2. *Suppose the set* $\{X : A_1 X + Y = U; A_2 X = e\} \cap (0,1)^{K \times N \times M}$ *is nonempty. Then Eq. 2.9 has a solution* $X^*(\mu) \in (0,1)^{K \times N \times M}$. *Also, there exists* $\mu^* > 0$ *such that for all* $\mu > \mu^*$, *the solution to Eq. 2.9 is unique.*

Proof. First, we show the existence of $X^*(\mu)$ for any $\mu > 0$. Define $S = [1/4, 3/4]^{M \times K \times N} \cap \{X : A_1 X + Y = U; A_2 X = e\}$. Since S is a compact set and $-f(X) + \mu \Phi(X,Y)$ is a continuous function on S, there exist real numbers L_1 and L_2 such that $L_1 \leq -f(X) + \mu \Phi(X,Y) \leq L_2$ for all S. Since $-f(X) + \mu \Phi(X,Y) \to \infty$ as $X_j^{k,i} \to 0^+$ or 1^-, there exists ε such that for all $X \in ((0,\varepsilon] \cup [1 - \varepsilon, 1)^{M \times K \times N} \cap \{X : A_1 X + Y = U; A_2 X = e\}$,

$$-f(X) + \mu \Phi(X,Y) > L_2. \tag{2.10}$$

Furthermore, ε must be greater than $1/4$. Define $S_1 = [\varepsilon, 1 - \varepsilon]^{M \times K \times N} \cap \{X : A_1 X + Y = U; A_2 X = e\}$. Again, by continuity of $-f(X) + \mu \Phi(X,Y)$ on the compact set S_1, there exists $Z \in S_1$ such that $-f(Z) + \mu \Phi(Z,Y) \leq -f(X) + \mu \Phi(X,Y)$ for all $X \in S_1$. Moreover, $-f(Z) + \mu \Phi(Z,Y) \leq L_2$ as $S \subset S_1$. By utilizing Eq. 2.10, we have $-f(Z) + \mu \Phi(Z,Y) < -f(X) + \mu \Phi(X,Y)$ for all $X \in (0,1)^{M \times K \times N} \setminus S_1$. Thus, Z is required for $X^*(\mu)$.

The uniqueness of $X^*(\mu)$ for sufficiently large μ follows from Lemma 3.1 of [17] and the convexity of the feasible region in Eq. 2.9.

Observe that the unique solution is obtained for Eq. 2.9 with no definitive rounding, i.e., the variables are not close to 0 or 1. Motivated by this, we present a plenty term, which is defined by

$$\gamma \sum_{j=1}^{M} \sum_{k=1}^{K} \sum_{i=1}^{N} X_j^{k,i} \left(1 - X_j^{k,i} \right),$$

where $\gamma > 0$ is a penalty parameter. Notice that the penalty term forces binary variables to their bounds. Therefore, the actual problem we are solving is

$$\min \quad F(X,Y) \tag{2.11}$$

$$\text{s.t.} \quad A_1 X + Y = U, \tag{2.12}$$

$$A_2 X = e, \tag{2.13}$$

where

$$F(X,Y) = -f(X) + \mu \Phi(X,Y) +$$

$$\gamma \sum_{j=1}^{M} \sum_{k=1}^{K} \sum_{i=1}^{N} X_j^{k,i} \left(1 - X_j^{k,i} \right).$$

The following theorem demonstrates that Eqs. 2.9 and 2.11 share the same solution.

Theorem 3. *There exists $\gamma^* > 0$ such that for all $\gamma > \gamma^*$, Eqs. 2.9 and 2.11 have the same minimizer.*

Proof. First, let $\Theta(X) = -f(X) + \mu \Phi(X,Y)$; then, let $t^{(p)}$ denote the set elements of $\{0,1\}^{M \times K \times N}$, and $T_{(p)}$ denote the set $\{\dot{X} \in [0,1]^{M \times K \times K} : \|X - t^{(p)}\| < 1/4\}$. Suppose $X \in T_{(q)}$ for some q. Then for indices i,j,k in which $t_j^{k,i(q)} = 0$, we have $x_j^{k,i} = |x_j^{k,i} - t_j^{k,i(q)}| \le \|X - t^{(q)}\| \le 1/4$, so that

$$|x_j^{k,i} - t_j^{k,i(q)}| = x_j^{k,i} \le 2x_j^{k,i}(1 - x_j^{k,i}). \tag{2.14}$$

Similarly, for indices i,j,k in which $t_j^{k,i(q)} = 1$, we have $1 - x_j^{k,i} = |x_j^{k,i} - t_j^{k,i(q)}| \le \|X - t^{(q)}\| \le 1/4$, so that

$$|x_j^{k,i} - t_j^{k,i(q)}| = 1 - x_j^{k,i} \le 2x_j^{k,i}(1 - x_j^{k,i}). \tag{2.15}$$

By Taylor's theorem, there exists $\tau \in [0,1]^{M \times K \times N}$ such that $\Theta(X) = \Theta(t^{(q)}) + (\nabla\Theta(\tau))^T (X - t^{(q)})$. Since $\nabla\Theta(t^{(q)})$ is continuous on the compact set $[0,1]^{M \times K \times N}$, there exists some constant $L_0 > 0$ such that

$$\Theta(t^{(q)}) - \Theta(X) \leq |\Theta(X) - \Theta(t^{(q)})|$$

$$= |(\nabla\Theta(\tau))^T (X - t^{(q)})|$$

$$\leq L_0 \|(X - t^{(q)})\|$$

$$= L_0 \sqrt{\sum_{j=1}^{M} \sum_{k=1}^{K} \sum_{i=1}^{N} (x_j^{k,i} - t_j^{k,i(q)})^2}$$

$$\leq L_0 \sum_{j=1}^{M} \sum_{k=1}^{K} \sum_{i=1}^{N} |x_j^{k,i} - t_j^{k,i(q)}|$$

from Eqs. 2.14 and 2.15

$$\leq 2L_0 \sum_{j=1}^{M} \sum_{k=1}^{K} \sum_{i=1}^{N} x_j^{k,i}(1 - x_j^{k,i}).$$

So, if $\gamma > 2L_0$,

$$\Theta(t^{(q)}) \leq \Theta(X) + \gamma \sum_{j=1}^{M} \sum_{k=1}^{K} \sum_{i=1}^{N} x_j^{k,i}(1 - x_j^{k,i})$$

for $X \in T_{(q)}$.

Suppose $X \in C$, where $C = [0,1]^{M \times K \times N} \setminus (\bigcup T_{(p)})$. By the continuity of $\sum_{j=1}^{M} \sum_{k=1}^{K} \sum_{i=1}^{N} x_j^{k,i}(1 - x_j^{k,i})$ on the compact set C, there exist L_1 and L_2 such that $\Theta(X) \geq L_1$ and $\sum_{j=1}^{M} \sum_{k=1}^{K} \sum_{i=1}^{N} x_j^{k,i}(1 - x_j^{k,i}) \geq L_2$ for all $X \in C$. In particular, $L_2 > 0$ since $X \neq t^{(p)}$ for all p. This implies that for all $X \in C$,

$$\Theta(X) + \gamma \sum_{j=1}^{M} \sum_{k=1}^{K} \sum_{i=1}^{N} x_j^{k,i}(1 - x_j^{k,i}) \geq L_1 + \gamma L_2$$

$$\geq \Theta(t^{(p)})$$

for all p if $\gamma \geq (L_3 - L_1)/L_2$, where $L_3 = \max_p \Theta(t^{(p)})$. Thus, if $\gamma > \max\{2L_0, (L_3 - L_1)/L_2\}$, we have $L_3 \leq \Theta(X) + \gamma \sum_{j=1}^{M} \sum_{k=1}^{K} \sum_{i=1}^{N} x_j^{k,i}(1 - x_j^{k,i})$. Letting

$$p' = \underset{p:A_1 t^{(p)} + Y = U; A_2 t^{(p)} = e}{\arg\min} \Theta(t^{(p)})$$

and $\gamma^* = \max\{2L_0, (L_3 - L_1)/L_2\}$, we also have

$$\Theta(t^{(p')}) + \gamma \sum_{j=1}^{M} \sum_{k=1}^{K} \sum_{i=1}^{N} t_j^{k,i(p')}(1 - t_j^{k,i(p')})$$

$$= \Theta(t^{(p')})$$

$$\geq \Theta(X) + \gamma \sum_{j=1}^{M} \sum_{k=1}^{K} \sum_{i=1}^{N} x_j^{k,i}(1 - x_j^{k,i})$$

for all $X \in [0,1]^{M \times K \times N} \cap \{p : A_1 X + Y = U; A_2 X = e\}$ if $\gamma > \gamma^*$. The theorem follows from the observation that $t^{(p')}$ is the minimizer of Eqs. 2.9 and 2.11.

The following theorem indicates that the solution is obtained for Eq. 2.11 with variables close to 0 or 1 by adding the plenty term.

Theorem 4. *Let* $X(\gamma, \mu)$ *be any local minimizer of Eq. 2.11. Then* $\lim_{\gamma \to \infty} \lim_{\mu \to 0} X_j^{k,i}$ $(\gamma, \mu) = 0$ *or 1,* $\forall j, \forall k, \forall i$.

Proof. Let $\gamma > 0$. The objective function in Eq. 2.11 can be rewritten as

$$F(X,Y) = f_\gamma(X,Y) + \mu \Phi(X,Y),$$

where

$$f_\gamma(X,Y) = -f(X) + \gamma \sum_{j=1}^{M} \sum_{k=1}^{K} \sum_{i=1}^{N} X_j^{k,i}\left(1 - X_j^{k,i}\right).$$

Also, let $X(\gamma)$ be a solution to

$$\min \quad f_\gamma(X,Y)$$
$$\text{s.t. } A_1 X + Y = U,$$
$$A_2 X = e,$$
$$0 \leq X \leq e.$$

Notice that

$$\lim_{\mu \to 0} X(\gamma, \mu) = X(\gamma).$$

Observe that the above problem becomes a sequence of penalty subproblems for Eq. 2.8 when γ is used as the penalty parameter. By Theorem 3, we know that $X_j^{k,i}(\gamma, \mu) = \{0, 1\}$ for sufficiently large γ so that $X_j^{k,i}(\gamma) \to 0$ or 1 as $\gamma \to \infty$, $\forall i, \forall j, \forall k$. Finally, this implies $\lim_{\gamma \to \infty} \lim_{\mu \to 0} X_j^{k,i}(\gamma, \mu) = 0$ or 1, $\forall j, \forall k, \forall i$.

2.2.2   Request-Allocation Algorithm

The first-order optimality conditions of Eq. 2.11 can be written as

$$\nabla_X F(X,Y) + A_1^T \alpha + A_2^T \beta = 0, \tag{2.16}$$

$$\nabla_Y F(X,Y) + \alpha = 0, \tag{2.17}$$

$$A_1 X + Y = U, \tag{2.18}$$

$$A_2 X = e, \tag{2.19}$$

where $\alpha = (\alpha_1, \alpha_2, \dots, \alpha_M)^T$ corresponds to the Lagrange multiplier for the constraint in Eq. 2.12. Similarly, $\beta = (\beta_{1,1}, \dots, \beta_{1,N}, \dots, \beta_{K,1}, \dots, \beta_{K,N})$, is a $(KN \times 1)$ matrix that denotes the Lagrange multiplier for the constraint in Eq. 2.13. Directly applying Newton's method yields

$$\begin{bmatrix} F_1 & F_2 & A_1^T & A_2^T \\ F_3 & F_4 & I & 0 \\ A_1 & I & 0 & 0 \\ A_2 & 0 & 0 & 0 \end{bmatrix} \begin{bmatrix} \Delta X \\ \Delta Y \\ \Delta \alpha \\ \Delta \beta \end{bmatrix} = \begin{bmatrix} r_1 \\ r_2 \\ r_3 \\ r_4 \end{bmatrix}, \tag{2.20}$$

where $F_1, F_2, F_3,$ and F_4 are as follows:

$$F_1 = \nabla_{XX} F(X,Y),$$

$$F_2 = \nabla_{XY} F(X,Y),$$

$$F_3 = \nabla_{YX} F(X,Y),$$

$$F_4 = \nabla_{YY} F(X,Y).$$

Furthermore, $r_1, r_2, r_3,$ and r_4 are defined by

$$r_1 = -\left(\nabla_X F(X,Y) + A_1^T \alpha + A_2^T \beta\right),$$

$$r_2 = -\left(\nabla_Y F(X,Y) + \alpha\right),$$

$$r_3 = -\left(A_1 X + Y - U\right),$$

$$r_4 = -\left(A_2 X - e\right).$$

Observe that Eq. 2.20 can be reduced as follows:

$$\begin{bmatrix} H & A_2^T \\ A_2 & 0 \end{bmatrix} \begin{bmatrix} \Delta X \\ \Delta \beta \end{bmatrix} = \begin{bmatrix} u \\ r_4 \end{bmatrix}, \tag{2.21}$$

where

$$H = F_1 - F_2 A_1 - A_1^T F_3 + A_1^T F_4 A_1,$$

$$u = r_1 - F_2 r_3 - A_1^T r_2 + A_1^T F_4 r_3.$$

The rest of the Newton directions can be obtained using the following equations:

$$\Delta Y = r_3 - A_1 \Delta X,$$

$$\Delta \alpha = r_2 - F_3 \Delta X - F_4 \Delta Y.$$

Letting Z be a matrix with columns that form a basis null-space matrix of A_2; then, we have $A_2 Z = 0$. Let X_0 be any feasible point such that $A_2 X_0 = e$; thus, $\Delta X = Zx$ for some x. Substituting this into the top part of Eq. 2.21 and premultiplying both sides by Z^T yields

$$Z^T H Z x = Z^T u, \qquad (2.22)$$

where x can be obtained using a conjugate gradient method [18]. After obtaining the corresponding ΔX, we then obtain ΔY, $\Delta \alpha$, and $\Delta \beta$. By performing a linear search method to determine the step size ρ_l, we set $X^{l+1} = X^l + \rho_l \Delta X$. Similarly, we update Y^{l+1}, α^{l+1} and β^{l+1}. Our request allocation algorithm based on logarithmic smoothing is summarized in Algorithm 1. The following theorem states that our algorithm significantly converges to the unique solution.

Theorem 5. *Let $\{\mu_t\}_{t=1}^{\infty}$ be a recursive sequence of positive numbers such that $\lim_{t \to \infty} \mu_t = 0$. Moreover, suppose that there exists $(X^*, Y^*, \alpha^*, \beta^*)$ that satisfies Eqs. 2.16–2.19. Then,*

$$\lim_{t \to \infty} X(\mu_t) = X^*.$$

Proof. Since $(X^*, Y^*, \alpha^*, \beta^*)$ satisfies Eqs. 2.16–2.19, the following equations hold:

$$\nabla_{X^*} F(X^*, Y^*) + A_1^T \alpha^* + A_2^T \beta^* = 0, \qquad (2.23)$$

$$\nabla_{Y^*} F(X^*, Y^*) + \alpha^* = 0, \qquad (2.24)$$

$$A_1 X^* + Y^* = U, \qquad (2.25)$$

$$A_2 X^* = e. \qquad (2.26)$$

Moreover, for each t,

$$\nabla_{X(\mu_t)} F(X(\mu_t), Y(\mu_t)) + A_1^T \alpha(\mu_t) + A_2^T \beta(\mu_t) = 0, \qquad (2.27)$$

$$\nabla_{Y(\mu_t)} F(X(\mu_t), Y(\mu_t)) + \alpha(\mu_t) = 0, \qquad (2.28)$$

$$A_1 X(\mu_t) + Y(\mu_t) = U, \qquad (2.29)$$

$$A_2 X(\mu_t) = e. \qquad (2.30)$$

Algorithm 1 Request-Allocation Algorithm Based on Logarithmic Smoothing

---

Input:

Request matrix, $[r_{k,i}]_{K \times N}$;

Bandwidth capacity, U_j;

Application instance placement matrix, $[w_{i,j}]_{N \times M}$;

Amount of bandwidth to handle one request when serving $a_i \in \mathcal{N}$, b_i;

Inherent cost, $\xi_{i,j}^k$;

Average response time, $\delta_{i,j}$;

Tolerance for function evaluation, ε;

Tolerance for barrier value, ε_μ;

Maximum value for penalty parameter, ε_γ;

Rounds of iteration, S;

Reduction ratio for barrier parameter, η_μ;

Reduction ratio for penalty parameter, η_γ;

Initial value of barrier parameter, μ_0;

Initial value of penalty parameter, γ_0;

Feasible starting point, r;

Value of the weight, λ_1;

Value of the weight, λ_2;

Output:

Request-allocation matrix, $\left[v_{i,j}^k\right]_{M \times N \times K}$;

1: Set $\gamma = \gamma_0$, $\mu = \mu_0$;

2: **while** $\gamma < \varepsilon_\gamma$ or $\mu > \varepsilon_\mu$ **do**

3:     Set $\left(X^0, Y^0, \alpha^0, \beta^0\right) = r$;

4:     set $l = 0$;

5:     **while** $l \leq S$ **do**

6:         **if** $F(X, Y) < \varepsilon\mu$ **then**

7:             Set $X^S = X^l$, $Y^S = Y^l$;

8:             Set $\alpha^S = \alpha^l$;

9:             Set $\beta^S = \beta^l$;

10:            Set $l = S$;

11:        **else**

12:            By applying the conjugate gradient method for Eq. 2.22, we obtain $\Delta X, \Delta Y, \Delta\alpha, \Delta\beta$;

13:            Perform a linesearch (see [19]) to determine ρ_l, set $X^{l+1} = X^l + \rho_l \Delta X$, and update $Y^{l+1}, \alpha^{l+1}, \beta^{l+1}$;

14:        **end if**

15:    **end while**

16:    Set $r = \left(X^S, Y^S, \alpha^S, \beta^S\right)$;

17:    Set $\mu = \eta_\mu \mu$;

18:    Set $\gamma = \frac{1}{\eta_\gamma} \gamma$;

19: **end while**

20: Obtain the request-allocation matrix by Eqs. 2.7 and 2.1;

21: **return** $\left[v_{i,j}^k\right]_{M \times N \times K}$;

From Eqs. 2.25, 2.26, 2.29, and 2.30, we obtain

$$\begin{bmatrix} A_1 & I \\ A_2 & 0 \end{bmatrix} \begin{bmatrix} X(\mu_t) - X^* \\ Y(\mu_t) - Y^* \end{bmatrix} = \begin{bmatrix} 0 \\ 0 \end{bmatrix}, \tag{2.31}$$

where I is an $(M \times M)$ identity matrix. Furthermore, from Eqs. 2.23, 2.24, 2.27, and 2.28, we have

$$\begin{bmatrix} H_1 \\ H_2 \end{bmatrix} + \begin{bmatrix} A_1 & I \\ A_2 & 0 \end{bmatrix}^T \begin{bmatrix} \alpha(\mu_t) - \alpha^* \\ \beta(\mu_t) - \beta^* \end{bmatrix} = \begin{bmatrix} 0 \\ 0 \end{bmatrix}, \tag{2.32}$$

where $H_1 = \nabla_{X(\mu_t)} F(X(\mu_t) - \nabla_{X^*} F(X^*, Y^*)$ and $H_2 = \nabla_{Y(\mu_t)} F(X(\mu_t) - \nabla_{Y^*} F(X^*, Y^*)$. Premultiplying both sides of Eq. 2.32 by $\begin{bmatrix} X(\mu_t) - X^* \\ Y(\mu_t) - Y^* \end{bmatrix}^T$ and using Eqs. 2.25, 2.29, and 2.31 yields

$$(X(\mu_t) - X^*)^T (H_1 + H_2) = 0.$$

This implies that

$$\lim_{t \to \infty} X(\mu_t) = X^*.$$

2.3 Experiment Evaluation

2.3.1 Simulation Setup

We employ simulators and real-word traces to evaluate our request allocation algorithm. We simulate a provider that deploys three datacenters, marked as Datacenter 1 to 3, hosting three types of application instances altogether. We used Wikipedia request traces [20] as the real-word traces to represent the request traffic, which contains requests issued to Wikipedia from 04:10 a.m., September 19, 2007 GMT to 04:11 a.m., September 20, 2007 GMT. We assumed that the service provider periodically addresses the optimization problem, i.e., hourly. Figure 2.2 plots the hourly request traffic, where we extract three traces in all, denoted as Application 1 to 3, with each trace being for a 24 h duration.

Since the dataset does not contain any end-user information, we split the total requests among three users following a normal distribution. Without loss of generality, we assumed that each datacenter is equipped with the same set of application instances as well as the same fixed bandwidth capacity. Each type of application instance $a_i \in \mathcal{N}$ consumes the same amount of bandwidth, such as one unit of bandwidth when handling one corresponding request. We considered the

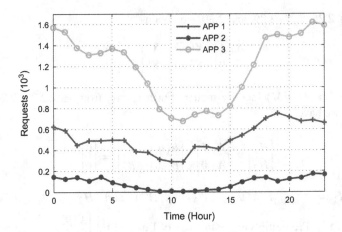

Fig. 2.2 Wikipedia request traces for 24 h

Table 2.1 Default simulation parameter values

Parameter	Value
M	3
K	3
N	3
$b_i, \forall i$	1
$w_j, \forall j$	$(1,1,1)$
$\lambda_1 = \lambda_2$	0.5

benefits of service providers and end-users to have the same importance; thus, we set $\lambda_1 = \lambda_2$. The default parameters are summarized in Table 2.1.

The inherent cost $\xi_{i,j}^k$ can be varied dynamically and periodically as end-users' demands change. For simplicity, in our simulation, we assumed that $\xi_{i,j}^k = 1$. Note that parameters $\theta_{i,j}$, $\sigma_{\tau_{i,j}}^2$, $\varphi_{i,j}$, and $\sigma_{\varphi_{i,j}}^2$ are all time varying as well as stochastic, requiring a prediction of their values at each interval. We obtained the prediction values of these variables for the current interval from the previous interval, i.e., the prediction values in hour 12 were based on information in hour 11. By using the prediction values of $\theta_{i,j}$, $\sigma_{\tau_{i,j}}^2$, $\varphi_{i,j}$, and $\sigma_{\varphi_{i,j}}^2$ for a given interval, the response time can be acquired.

2.3.2 Performance Analysis

We first evaluated the performance of our request allocation algorithm. The bandwidth capacity of each datacenter was set to 1000 units (i.e., $U_j = 1000$ for all j). Figure 2.3 shows the average bandwidth utilization with total requests for a 24 h period of time. Clearly, the average bandwidth utilization figure closely follows the

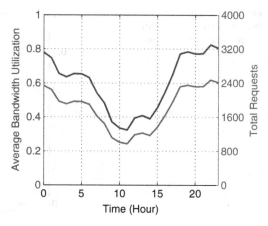

Fig. 2.3 Average bandwidth utilization, $U_j = 1000$

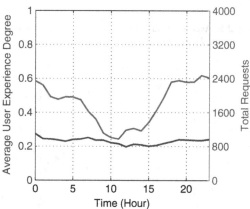

Fig. 2.4 Average user experience degree, $U_j = 1000$

total requests. We observed that the average bandwidth utilization is below 0.8 most of the time. This allows providers ample residual financial bandwidth resources to do other work without having to consider the extra bandwidth cost involved in this request allocation problem.

The performance can be better explained by Fig. 2.4, which shows the average user experience degree versus total requests. We found that average user experience maintains a relatively stable value of 0.24, which demonstrates that our request allocation algorithm guarantees stable user experience with no relation to total requests. Although requests are picked, service providers can offer stable service to end-users.

To understand the performance of our request allocation algorithm on a microscopic level, we plot the cumulative distribution function (CDF) per request across all end-users and all hours in Fig. 2.5. Observe that most of the requests, more than 95 %, are served with a response time less than 1000 ms. Only less than 5 % are more than 1000 ms, which we will implement in our future work.

Fig. 2.5 Cumulative
distribution function (CDF)
of per request response time,
$U_j = 1000$

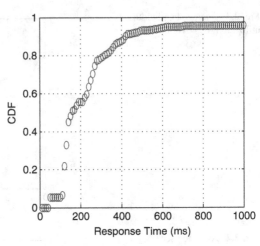

Fig. 2.6 Average bandwidth
utilization, $U_j = 900$

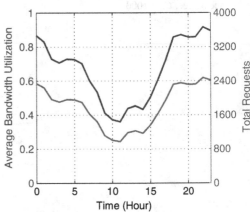

We also considered the performance of our request algorithm with a reduced bandwidth capacity (i.e., $U_j = 900$ for all j) while keeping all other settings unchanged. Compared to Fig. 2.3, Fig. 2.6 demonstrates higher average bandwidth utilization. Figure 2.7 also shows that the average user experience degree is unchanged compared to that in Fig. 2.4. This further verifies that stable service is guaranteed. As shown in Fig. 2.8, the response time per request is similar to that in Fig. 2.5. All these facts show that when holding total requests, providers can reduce the bandwidth capacity for each datacenter without reducing the user experience degree, which can indirectly improve provider profits.

Now, we investigate the fairness achieved with our request allocation algorithm. We used a simple model that shares all of the same constraints with our model, but uses

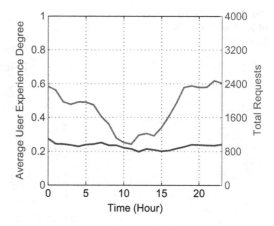

Fig. 2.7 Average user experience degree, $U_j = 900$

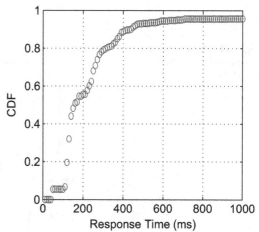

Fig. 2.8 CDF of per request response time, $U_j = 900$

$$\max \begin{pmatrix} \sum\limits_{j=1}^{M} P_j \\ \sum\limits_{k=1}^{K} T_k \end{pmatrix}$$

as the objective function instead of Eq. 2.2. For simplicity, we denote this model by RAUF, and denote our model by RAF. Note that RAUF does not contain the fairness issue and can also be solved by logarithmic smoothing, and then, Newton's method, the conjugate method, and linesearch method can be applied. We determined that RAUF always allocates requests to a datacenter with lower response time for the objective that maximizes the sum of the total bandwidth utilization and total user experience degree; thus, user experience can be guaranteed. We evaluated the fairness for different bandwidth capacities, 1000 and 900, for each datacenter, respectively.

Fig. 2.9 Standard deviation of bandwidth utilization, $U_j = 1000$

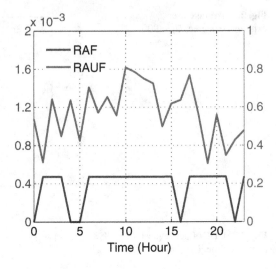

Fig. 2.10 Standard deviation of bandwidth utilization, $U_j = 900$

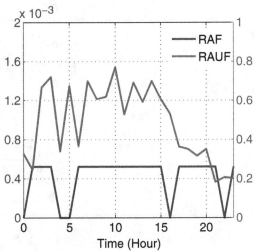

Figure 2.9 first shows the fairness comparison of RAF and the simplified RAUF, where we use the standard deviation of bandwidth utilization for each datacenter at each time as the fairness measure. A fair request allocation strategy can achieve a load balance between datacenters with a smaller standard deviation, while a poor algorithm that does not consider fairness has a larger standard deviation. Observe from Fig. 2.9 that in a 24 h period, RAF always achieves a much smaller standard deviation, which translates to a much better load balance among datacenters. The load balance generally improves over the time, which verifies the effectiveness of our model. This is further confirmed in Fig. 2.10 with a reduced bandwidth capacity (i.e., $U_j = 900$ for all j) for each datacenter.

Fig. 2.11 Standard deviation of the user experience degree, $U_j = 1000$

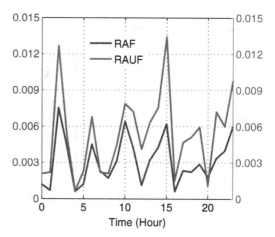

Fig. 2.12 Standard deviation of the user experience degree, $U_j = 900$

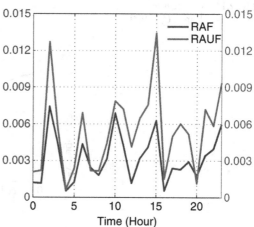

We also studied fairness between end-users. Figure 2.11 shows the standard deviation of the average user experience degree for RAF and RAUF. Observe that RAF achieves a much smaller standard deviation than RAUF, which implies RAF achieves better fairness for users than RAUF. This point is further verified in Fig. 2.12, where the bandwidth capacity for each datacenter reduces to 900. Figures 2.13 and 2.14 present the average bandwidth utilization for different bandwidth capacities. Although RAUF provides better user experience than RAF, its fairness for end-users is worse than that of RAF. Overall, we believe that our algorithm is practical for real-word problems.

Fig. 2.13 Average user
experience degree, $U_j = 1000$

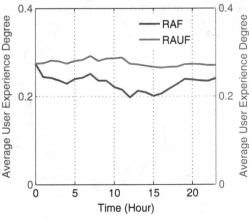

Fig. 2.14 Average user
experience degree, $U_j = 900$

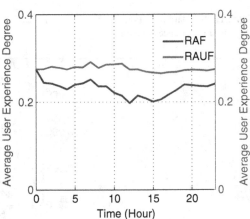

2.4   Conclusion

In this chapter, we took advantage of the central control provided by SDN to address
the request allocation problem in distributed datacenters. Unlike existing work on
request allocation, we developed a general formulation of the request allocation
problem considering both the benefit of service providers and end-users. We applied
two Nash bargaining games, where each Nash product denoted the benefit of the
service provider and the benefit of end-users, respectively. Specifically, we proposed
the concept of the user experience degree to capture user experience, which depends
on the most important metric response time inside the datacenter and some inherent
cost. Furthermore, we presented an efficient request allocation algorithm based on
logarithmic smoothing. After applying Newton's method, the conjugate method, and
linear search method, we theoretically proved that our request allocation algorithm
significantly converges to a unique solution. We evaluated its efficiency and
practicality using real-word traces in our simulations. Our algorithm simultaneously

achieved high bandwidth utilization as well as load balance for service providers and good user experience, as well as fairness for end-users. Extensive simulations showed that our algorithm exhibits better fairness than a framework where fairness is not considered. Overall, our work shows that it is feasible for an SDN controller to implement request allocation.

References

1. C. Guo, H. Wu, K. Tan, et al. Dcell: A Scalable and Fault-tolerant Network Structure for Data Centers. ACM SIGCOMM Computer Communication Review, 2008, 38(4): 75–86.
2. A. Singh, M. Korupolu and D. Mohapatra. Server-storage Virtualization: Integration and Load Balancing in Data Centers. Proceedings of the 2008 ACM/IEEE conference on Supercomputing. ACM/IEEE, 2008: 1–12.
3. R. Buyya, R. Ranjan and R. N. Calheiros. Intercloud: Utility-oriented Federation of Cloud Computing Environments for Scaling of Application Services. Algorithms and Architectures for Parallel Processing. Springer Berlin Heidelberg, 2010: 13–31.
4. A. Qureshi, R. Weber, H. Balakrishnan, et al. Cutting the Electric Bill for Internet-scale Systems. ACM SIGCOMM Computer Communication Review, 2009, 39(4): 123–134.
5. P. X. Gao, A. R. Curtis, B. Wong, et al. It's Not Easy Being Green. ACM SIGCOMM Computer Communication Review, 2012, 42(4): 211–222.
6. Z. Liu, M. Lin, A. Wierman, et al. Greening Geographical Load Balancing. Proceedings of the ACM SIGMETRICS Joint International Conference on Measurement and Modeling of Computer Systems. ACM, 2011: 233–244.
7. H. Xu and B. Li. Cost Efficient Datacenter Selection for Cloud Services. Proceedings of the 1st IEEE International Conference on Communications in China (ICCC 2012). IEEE, 2012: 51–56.
8. K. Boloor, R. Chirkova, Y. Viniotis, and T. Salo. Dynamic Request Allocation and Scheduling for Context Aware Applications Subject to A Percentile Response Time SLA in A Distributed Cloud. Proceedings of the 2nd International Conference on Cloud Computing Technology and Science (CloudCom). IEEE, 2010: 464–472.
9. P. Wendell, J. W. Jiang, M. J. Freedman, et al. Donar: Decentralized Server Selection for Cloud Services. ACM SIGCOMM Computer Communication Review, 2010, 40(4): 231–242.
10. H. Xu and B. Li. Joint Request Mapping and Response Routing for Geo-distributed Cloud Services. Proceedings of the IEEE INFOCOM. IEEE, 2013: 854–862.
11. S. Shenker, M. Casado, T. Koponen, et al. The Future of Networking, and the Past of Protocols. Open Networking Summit, Stanford University, USA, October 2011.
12. W. Li, H. Qi, K. Li, et al. Joint Optimization of Bandwidth for Provider and Delay for User in Software Defined Data Centers. IEEE Transactions on Cloud Computing. DOI: 10.1109/TCC.2015.2402677.
13. E. Cela. The Quadratic Assignment Problem: Theory and Algorithms. Kluwer Academic, Dordrecht, 1998.
14. A. Muthoo. Bargaining Theory with Applications. Cambridge University Press, 1999.
15. G. Bolch, S. Greiner, H. Meer, et al. Queueing Networks and Markov Chains: Modeling and Performance Evaluation with Computer Science Applications. John Wiley & Sons, 2006.
16. W. Murray, K. Ng. An Algorithm for Nonlinear Optimization Problems with Binary Variables. Computational Optimization and Applications, 2010, 47(2): 257–288.
17. K. M. Ng. A Continuous Approach for Solving Nonlinear Optimization Problems with Discrete Variable. Stanford: Department of Management Science and Engineering of Stanford University, 2002.

18. G. H. Golub and C. F. Loan. Matrix Computation, The John Hopkins University Press, Baltimore and London, 1996.
19. A. Forsgren and W. Murray. Newton methods for large-scale linear equality-constrained minimization, SIAM Journal on Matrix Analysis and Applications, 1993, 14(2): 560–587.
20. Wikipedia Request Traces, http://www.wikibench.eu/.

Chapter 3
Software-Defined Networking Controller Placement in Distributed Datacenters

Abstract From the previous chapter, one can see that software-defined networking (SDN) can be applied to large-scale distributed datacenter networks for network control and management. When this occurs, a logically centralized and physically distributed control plane is usually required. This type of control plane consists of multiple controllers communicating with one another. The collaboration of these controllers facilitates the maintenance of a global consistent view of the entire network. However, there are many new problems that must be addressed when a distributed control plane is deployed in a large-scale network; in particular, controller placement is one key problem. Controller placement refers to selecting the proper positions of the controllers to further improve the scalability and performance of the distributed control plane. In this chapter, we propose a novel placement metric for deploying multiple controllers that measures the cost when controllers with limited capacity handle request messages from switches. Then, we formulate the optimal controller placement problem as an integer linear program (ILP) and use an effective approximation algorithm to find its solution. We conduct intensive experiments based on many real topologies. Our results demonstrate that our strategy can significantly improve performance over existing strategies in terms of both cost and load balance.

In recent years, the use of software-defined networking (SDN) has increased and deepened. Using a well-defined application programming interface (API) provided by an SDN controller, routing can be customized for network control, and a network-wide view for network management can be maintained. Although we can benefit from the centralized control realized by SDN, it is big challenge to deploy SDN in large-scale networks such as distributed datacenter networks and wide area networks (WANs). When adopting a single controller in large-scale networks, problems arise such as limited scalability or a single point of failure. To address these problems, a logically centralized and physically distributed control plane is proposed. This type of control plane usually consists of multiple controllers. These controllers communicate with each other to realize centralized control. For example, ON.Lab launched one distributed SDN control platform named the Open Network Operating System (ONOS) [1].

© The Author(s) 2016

H. Qi, K. Li, *Software-Defined Networking Applications*
in Distributed Datacenters, SpringerBriefs in Electrical and Computer Engineering,
DOI 10.1007/978-3-319-33135-5_3

When deploying a distributed SDN control platform in large-scale distributed datacenter networks, the placement of multiple controllers is a key problem. Controller placement refers to selecting the positions of multiple controllers in a given network. More and more users have discovered that controller placement strategies influence every aspect of an SDN, from node-to-controller latencies, network availability, to performance metrics. Therefore, little research has been conducted on controller placement. Zhang et al. focused on controller placement in a split architecture network, where the control platform consists of a set of commodity servers connecting to one or more switches [2]. In their paper, they proposed a min-cut-based graph partitioning algorithm for controller placement, minimizing the likelihood of loss of connectivity between controllers and switches. Heller et al. addressed the controller placement problem with the objective of minimizing the latencies between switches and controllers [3]. In particular, they examined the impact of controller placement on average and worst-case latencies with real topologies. Bari et al. introduced a dynamic controller provisioning problem (DCPP) [4]. Moreover, they presented a framework that can adapt the number of controllers and their corresponding locations by changing network conditions. Hu et al. addressed the controller placement problem in order to maximize the reliability of control networks [5]. To achieve their goal, they presented a metric to measure the reliability of an SDN controller. Guo et al. studied the controller placement problem for network resilience [6]. To improve the resilience of SDN, they designed a novel metric for resilience and a new solution for controller placement. Muller et al. proposed a controller placement strategy named Survivor [7]. In the Survivor strategy, path diversity, capacity, and failover mechanisms are all taken into consideration. Lange et al. presented a framework for Pareto-based optimal controller placement [8]. In this framework, various important metrics were considered including the latency from nodes to controllers, the latency among controllers, resilience against nodes, and link failures.

These existing controller placement strategies, however, do not take the limited-capacity of controllers into consideration. In fact, a single controller usually cannot handle a large number of request messages originating from all infrastructure switches. Another major problem of these strategies is that the load of a controller should be balanced. As shown in Fig. 3.1, where different markers and colors illustrate switch-to-controller assignments, Controller A controls four switches, but Controller B controls 10 switches with a much higher load. Some controllers control too many switches, exceeding their limited capacities.

These problems motivate us to answer the following question: given a physical network and number of controllers, how should these controllers be placed so that the overall cost of installing forwarding rules from capacity-constrained SDN controllers to switches is minimized? To address this problem, we propose a new solution for optimal controller placement. In this solution, we first define the cost and formulate optimal controller placement as an integer linear programming (ILP) problem. We then develop a $(3 + 1/|S|)$-approximation algorithm using Lagrangian relaxation to find its solution. Through extensive simulations over real topologies, we show that the proposed approximation algorithm achieves both cost efficiency and load balance for controller placement.

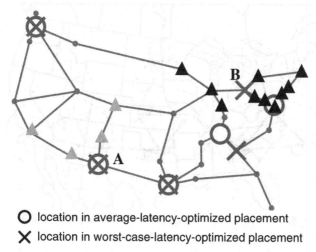

○ location in average-latency-optimized placement

✕ location in worst-case-latency-optimized placement

Fig. 3.1 Load-imbalanced examples

The remainder of this chapter is organized as follows. Section 3.1 defines our novel placement metric and formulates the controller placement problem. Section 3.2 presents a different approach based on one existing approximation algorithm and analyzes its approximation ratio. Section 3.3 discusses our experimental results. Finally, Sect. 3.4 concludes this chapter.

3.1 Placement Problem of Multiple Software-Defined Networking Controllers

In this section, we define our novel placement metric and formulate the controller placement problem as an ILP problem.

3.1.1 System Model

We consider an SDN-based distributed datacenter network consisting of OpenFlow-enabled switches and multiple capacity-limited controllers, where each controller maintains a global view of the network. Generally, a switch is controlled by at least one controller. The abstract view of the physical network with multiple controllers is shown in Fig. 3.2, where S is the set of switches and $C(C \subseteq S)$ is the set of locations where a controller can be deployed.

When a new flow arrives at a switch, the switch first checks its forwarding table for a matching entry. If a matching forwarding rule exists, packets in the flow are

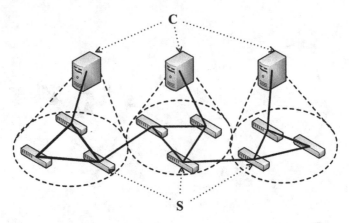

Fig. 3.2 Abstract view of a physical software-defined networking (SDN) network with multiple controllers

Table 3.1 Notation and definitions

Notation	Definition
S	The set of switches, $|S| = n$
C	The set of controllers
k	The maximum number of controllers
d_{ij}	The cost along the shortest path between switch i and controller j
u_j	The maximum number of requests controller j can handle in a unit of time
δ	The maximum allowable cost of any adopted control path
f_i	The number of flows from switch i in a unit of time
y_j	A binary variable indicating whether controller j is active
x_{ij}	A binary variable indicating whether switch i is controlled by controller j

forwarded according to the matching rule. Otherwise, the switch sends a *packet-in* OpenFlow message to its controller. The controller then computes a path with its global network view and installs forwarding rules to all switches along the path of the flow.

Communications between switches and controllers are made by control paths, each of which is constituted by existing connections between switches in the network. Let d_{ij} denote the cost of the shortest path between switch i and controller j expressed in terms of the propagation delay. Any adopted control path should have cost at most δ (expressed in the same unit as d_{ij}).

The associated cost is then considered as the metric for the controller placement strategies, which can be expressed by $D = \sum_{i \in S} \sum_{j \in C} f_i x_{ij} d_{ij}$, where f_i represents the number of flows from switch i, and x_{ij} is an indicator variable denoting whether switch i is controlled by controller j. The goal is to find a placement of at most k controllers such that the resulting cost D is minimized. The important notation used throughout this chapter is listed in Table 3.1.

3.1.2 Problem Formulation

Based on the above system model, the controller placement problem studied in this chapter can be formulated as follows:

$$\min \quad \sum_{i \in S} \sum_{j \in C} f_i x_{ij} d_{ij} \tag{LP1}$$

$$\text{s.t.} \quad \sum_{j \in C} x_{ij} \geq 1, \qquad \forall i \in S \tag{3.1}$$

$$\sum_{j \in C} y_j \leq k, \tag{3.2}$$

$$\sum_{i \in S} x_{ij} f_i \leq y_j u_j, \quad \forall j \in C \tag{3.3}$$

$$x_{ij} d_{ij} \leq \delta, \qquad \forall i \in S, \forall j \in C \tag{3.4}$$

$$x_{ij} \leq y_j, \qquad \forall i \in S, \forall j \in C \tag{3.5}$$

$$x_{ij} \in \{0, 1\}, \qquad \forall i \in S, \forall j \in C \tag{3.6}$$

$$y_j \in \{0, 1\}, \qquad \forall j \in C \tag{3.7}$$

Equation 3.1 guarantees that every switch is controlled by at least one controller at any time. Let y_j be a binary variable indicating that a controller will be deployed at j if y_j is equal to 1; otherwise, $y_j = 0$. Equation 3.2 indicates that at most k controllers can be deployed. Recall that each controller j has limited processing capacity denoted by u_j. Equation 3.3 ensures that a controller must satisfy the requests from the switches assigned to it. Equation 3.4 indicates that any adopted control path should be of cost at most δ. Finally, Eq. 3.5 enforces a controller j must be active, i.e., $y_j = 1$, if any switch i allocates to it, i.e., $x_{ij} = 1$.

3.2 Efficient Controller Placement Approximation Algorithm

Our controller placement problem is NP-hard because it is a generalized minimum k-median problem [9], which is one of the most widely studied NP-hard problems in location theory. While a number of algorithms with constant approximation ratios have been proposed for the classic minimum k-median problem, none of them can be directly applied to solve this controller placement problem. Instead, we proposed a different approach based on one existing approximation algorithm [10] where the main algorithmic idea is a new extension of the primal–dual schema, and its degree of approximation is three for the metric uncapacitated facility location problem.

Specifically, we model the network as a bipartite graph $G(S \cup C, E)$, where any edge (i, j) in set E represents the shortest path from switch $i \in S$ to controller

Fig. 3.3 Bipartite graph abstract of the whole network

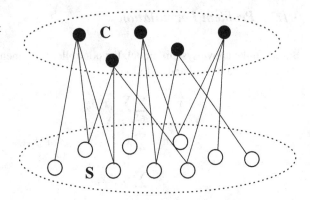

$j \in C$, as shown in Fig. 3.3. The goal of our controller placement optimization problem is to find the placement $\tilde{C}$ from the set of all possible controller placements such that $|\tilde{C}| \leq k$ and the incurred cost $D = \sum_{i \in S} \sum_{j \in C} f_i x_{ij} d_{ij}$ is minimized. In an effort to solve this NP-hard problem, we use the primal–dual schema to derive an approximation algorithm. The LP-relaxed formulation of the original problem and the corresponding dual program are given below by (LP2) and (LP3), respectively.

The LP-relaxation of this program is

$$\min \quad \sum_{i \in S} \sum_{j \in C} f_i x_{ij} d_{ij} \tag{LP2}$$

$$\text{s.t.} \quad \sum_{j \in C} x_{ij} \geq 1, \qquad \forall i \in S$$

$$\sum_{j \in C} -y_j \geq -k,$$

$$y_j u_j - \sum_{i \in S} x_{ij} f_i \geq 0, \quad \forall j \in C$$

$$-x_{ij} d_{ij} \geq -\delta, \qquad \forall i \in S, \forall j \in C$$

$$y_j - x_{ij} \geq 0, \qquad \forall i \in S, \forall j \in C$$

$$x_{ij} \geq 0, \qquad \forall i \in S, \forall j \in C$$

$$y_j \geq 0, \qquad \forall j \in C.$$

The dual program of this program is

$$\max \quad \sum_{i \in S} \alpha_i - \sum_{i \in S} \sum_{j \in C} \gamma_{ij} \delta - bk \tag{LP3}$$

$$\text{s.t.} \quad \alpha_i - \beta_{ij} \leq f_i d_{ij} + \gamma_{ij} d_{ij} + a f_i, \quad \forall i \in S, \forall j \in C$$

$$\sum_{i \in S} \beta_{ij} \leq b - a u_j, \qquad \forall j \in C.$$

$$\alpha_i \geq 0, \qquad\qquad\qquad \forall i \in S$$
$$\beta_{ij} \geq 0, \qquad\qquad\qquad \forall i \in S, \forall j \in C$$
$$\gamma_{ij} \geq 0, \qquad\qquad\qquad \forall i \in S, \forall j \in C$$
$$a, b \geq 0.$$

The algorithm is based on the primal–dual schema [10]. Before showing how this is done, several ideas are required. First, suppose (LP2) has an optimal solution that is an integral, say $\tilde{C} \subseteq C$ and $\Phi : S \to \tilde{C}$. Thus, under this solution, $y_j = 1$ if and only if $j \in \tilde{C}$ and $x_{ij} = 1$ if and only if $\Phi(i) = j$. The primal conditions are relaxed as follows: the switches are partitioned into two sets, *directly connected* and *indirectly connected*. β_{ij} is nonzero only if i is a directly connected switch, and $\Phi(i) = j$. For a directly connected switch i, $\alpha_i - \beta_{i\Phi(i)} = f_i d_{i\Phi(i)} + \gamma_{i\Phi(i)} d_{i\Phi(i)} + af_i$, and for each active controller j, $\sum_{i:\Phi(i)=j} \beta_{ij} = b - au_j$. Second, to solve (LP3), we use the Control Variate Method. Clearly, when $b = 0$, all controllers will be active in the algorithm. When b is very large, there will only be one active controller. The maximum value of b is ne_{max}, where n is the number of nodes and e_{max} is the length of the longest edge. Thus, the positive integers b and a satisfy $b \in [0, ne_{max}]$ and $a \in [0, b/u_j]$, respectively.

The algorithm consists of two phases. In the first phase, the algorithm finds a feasible dual solution and temporarily determines the controller placement C_t in a primal–dual fashion. In the second phase, a subset $\tilde{C}$ of C_t is chosen as the final solution by finding a mapping $\Phi(i) = j$ from switches to controllers in $\tilde{C}$. We now discuss these phases in greater detail.

Algorithm 1 shows the first phase of the primal–dual schema-based algorithm. Our goal is to find the largest dual solution. To achieve this target, the following ultimate process is required to deal with the non-covering packing pair of (LP1). Initially, every switch in the network is unconnected to the controller, the initial state of each switch is set to unconnected, and the dual variables of each switch i are set to 0, i.e., $\alpha_i = 0$ and $\beta_{ij} = 0$. Throughout this phase, for the dual variable α_i, the algorithm uniformly raises its value for each switch i until the state of the switch is changed. To maintain feasibility and satisfy the complementary slackness conditions, all other primal and dual variables request to change by the change in α_i.

For some edge (i, j) (edge (i, j) is the shortest path between switch i and controller j), when α_i is raised to $\alpha_i = f_i d_{ij} + \gamma_{ij} d_{ij} + af_i$, this edge is declared a *control path*. Similarly, the dual variable β_{ij} is increased uniformly. For controller $j \in C$, when β_{ij} is increased to $\sum_{i \in S} \beta_{ij} = b - au_j$, the controller is declared temporarily active. Therefore, if unconnected switches have a control path to this temporarily active controller, they are declared connected, and this controller is declared the *pseudo* controller. Later, when traversing all other unconnected switches, if switch i gets a control path to controller j, then switch i is also declared connected.

When the first phase is terminated, a switch may have been connected to several temporarily active controllers; we use C_t to denote the set of all such controllers. To ensure that a switch is only connected to one controller that will eventually be

---

Algorithm 1 The First Phase of the Primal–Dual Schema-Based Algorithm

---
Require:

   Set of switches, S;
   Set of controllers, C;
   Cost along shortest path between i and j, d_{ij};
   Capacity of controller j, u_j;
   Number of flows from switch i, f_i;
   Length of the longest edge, e_{max};
   Random positive integer, γ_{ij};

Ensure:

   The set of all temporarily active controllers, C_t;
 1: Initialize: Set $C_t = \varnothing$;
 2: Randomly select positive integer b in the interval $[0, ne_{max}]$;
 3: Randomly select positive integer a in the interval $[0, b/u_j]$;
 4: **for** each switch i **do**
 5:     Set dual variable $\alpha_i = 0$;
 6:     **while** switch i is unconnected to any controllers **do**
 7:         $\alpha_i ++$;
 8:     **end while**
 9:     **for** each edge (i,j) **do**
10:        Set dual variable $\beta_{ij} = 0$;
11:        **if** $\alpha_i = f_i d_{ij} + \gamma_{ij} d_{ij} + a f_i$ **then**
12:            edge (i,j) is a *control path*;
13:        **end if**
14:        **while** $\sum_{i \in S} \beta_{ij} < b - a u_j$ **do**
15:            $\beta_{ij} ++$;
16:        **end while**
17:        Declare controller j temporarily active;
18:        Add controller j to C_t;
19:     **end for**
20: **end for**
21: **return** C_t;

---

active, we choose a subset of temporarily active controllers in the second phase of the primal–dual schema-based algorithm shown in Algorithm 2.

In the second phase, let graph G_1 denote a subgraph of G. Initially, G_1 satisfies $G_1 = \varnothing$. If edge (i,j) is a control path of which the number of hops is at most two, it is added to G_1. All edges in G are traversed. Then, let $\tilde{G}$ denote the subgraph of G_1 induced on C_t. Our goal is to find any maximal independent set $\tilde{C} \subseteq \tilde{G}$ where all controllers in the set $\tilde{C}$ are active.

Let set F_i satisfy $F_i = \{j \in C | \beta_{ij} > 0\}$. Because $\tilde{C}$ is an independent set, there is at most one controller in F_i that is active. Traversing all switches, if controller $j \in F_i$ is active, then set the mapping from switches to $\tilde{C}$ as $\Phi(i) = j$, and directly connect switch i; otherwise, consider control path (i,j'). If controller j' is a *pseudo* controller and $j' \in \tilde{C}$, set $\Phi(i) = j'$ and directly connect switch i. If controller j' is a *pseudo* controller and $j' \notin \tilde{C}$, let controller $j \in \tilde{C}$ be any neighbor of j' in graph $\tilde{G}$, and set $\Phi(i) = j$ so that switch i is indirectly connected. Then, $\tilde{C}$ and Φ define a primal integral solution.

Algorithm 2 The Second Phase of the Primal–Dual Schema-Based Algorithm

Require:
    Set of all temporarily active controllers, C_t;
    Whole of network topology, G;
    Number of hops for edge (i,j), h_{ij};
Ensure:
    Maximal independent set of all active controllers, $\tilde{C}$;
    Mapping from switches to $\tilde{C}$, Φ;
1: Initialize: Set one subgraph of G to be $G_1 = \varnothing$;
2: **for** each edge (i,j) **do**
3:     **if** edge (i,j) is a *control path* **then**
4:         **if** $h_{ij} \leq 2$ **then**
5:             Add edge (i,j) to G_1;
6:         **end if**
7:     **end if**
8: **end for**
9: Set $\tilde{G}$ for the subgraph of G_1 induced on C_t;
10: Find any maximal independent set $\tilde{C} \subseteq \tilde{G}$;
11: **for** each switch i **do**
12:     Set $F_i =\{j \in C | \beta_{ij} > 0\}$;
13:     **if** $j \in F_i$ and j is active **then**
14:         Set the mapping $\Phi(i) = j$;
15:         **while** edge (i,j') is a *control path* **do**
16:             **if** $j' \in \tilde{C}$ **then**
17:                 Set the mapping $\Phi(i) = j'$;
18:             **else**
19:                 **if** j is any neighbor of j' **then**
20:                     Set $\Phi(i) = j$;
21:                 **end if**
22:             **end if**
23:         **end while**
24:     **end if**
25: **end for**
26: **return** $\tilde{C}$, Φ;

Finally, we obtain the following primal integral solution:

$$x_{ij} = 1 \quad \text{iff} \quad \Phi(i) = j, \tag{3.8}$$

and

$$y_j = 1 \quad \text{iff} \quad j \in \tilde{C}. \tag{3.9}$$

3.3   Experiment Evaluation

In this section, we evaluate the controller placement strategy based on realistic network topologies, including the Internet2 OS3E network [11] and the SINET4 network [12]. Internet2's Advanced Layer 2 Service has a large advantage in that

Table 3.2 Main characteristics of experiment networks

Networks	Main characteristics of experiment networks	
	Node number	Link number
OS3E	34	42
SINET4	75	77

it provides the research and education communities with effective and efficient, wide area 100 gigabit Ethernet technology. Not only do members have scalable and flexible global access to an open exchange network, but they can also build Layer 2 circuits (virtual local area networks) between endpoints on the Internet2 network. The service completely meets the wide-ranging needs of the research and education communities, both now and into the future. SINET4 is the biggest National Research and Education Network (NREN) in Japan, which connects approximately 700 universities and research institutes across Japan. Note that these topologies can be downloaded from the Internet Topology Zoo [13], which is a collection of annotated network graphs derived from public network maps. The key characteristics of these topologies are summarized in Table 3.2.

In our experiments, all nodes in the networks were capable of hosting controllers. We first evaluated the cost of a controller's metric discussed in Sect. 3.3 using the Internet2 OS3E topology. To characterize the cost of controllers against the number of controllers, we expanded our analysis to the SINET4 topology.

To investigate the cost of controllers on an SDN/OpenFlow WAN, we used Mininet 2.0 [14] to build analogies of the Internet2 OS3E's and the SINET4's network topologies and used POX [15] as a network controller with OpenFlow protocol version 1.0. Both Mininet and POX were running on Ubuntu 12.04.3 LTS. Finally, the networking devices were connected with 100 Mbps links, and the delay in each link was assigned a respective value, which is an acceptable latency between the controller and the switches.

3.3.1   Analysis of Internet2 OS3E

The brute-force method was adopted to find the values of $min \sum_{i \in S} \sum_{j \in C} f_i x_{ij} d_{ij}$ as follows: all possible locations of controllers and all possible connections between controllers and switches were measured and stored for analysis. For the OS3E topology, we used the results of the simulations to find the answer to the following question: given a physical network and number of controllers, how should these controllers be placed so that a pre-defined objective is optimized? Figure 3.4a shows the cumulative distribution function (CDF) of the cost obtained from all possible placements on OS3E when the number of controller varied from 1 to 5. The x-axis is the cost of controllers, and the y-axis shows the corresponding probability. Notice that the optimal values are obviously at the bottoms, and only a tiny percentage of placements are optimal.

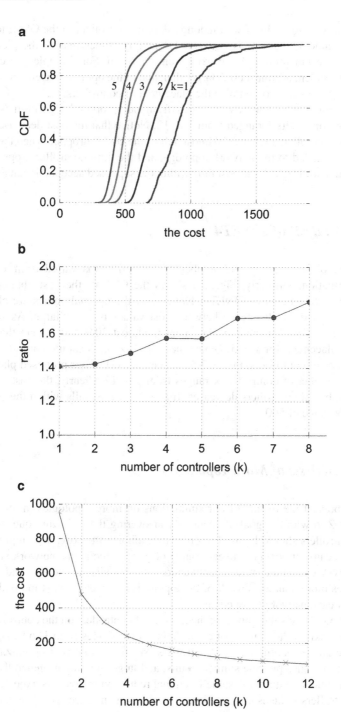

Fig. 3.4 Simulations results based on OS3E. (**a**) The cumulative distribution function (CDF) of the cost generated from all possible combinations of controller placements. (**b**) Ratio of randomly deploying to our proposed placement. (**c**) The cost for the number of controllers when $k = [1, 12]$

As shown in Fig. 3.4b, if we randomly deploy controllers to the OS3E topology for some value k, the cost is between 1.4 and 1.8 times that of the placements obtained from our proposed approximation algorithm. For example, the cost of a random placement is almost 60 % larger than that of the optimal placement for $k = 5$. Thus, it is worthwhile to optimize the placement of controllers.

Figure 3.4c shows the cost obtained from our proposed algorithm when the number of controllers k ranged from 1 to 12. Notice that the cost decreases as the number of controllers increases. Almost half of the cost is dropped when the second controller is added to the network topology, and the third controller drops the cost to less than half. If $k \geq 2$, the cost falls within the supposed acceptable range of 500.

3.3.2   Analysis of SINET4

For the SINET4 topology, the number of controller was varied from 1 to 6 in the investigation. Similarly, Fig. 3.5a shows the CDF of the cost obtained from all possible placements in SINET4. Notice that the optimal values are clearly at the bottoms, and the difference between cost values is rather large. As shown in Fig. 3.5b, the cost of a random placement is almost 70 % larger than that of our proposed placement for $k = 5$. Hence, network operators should carefully choose the locations of controllers. Figure 3.5c shows the cost of our proposed placements when the number of controllers k ranges from 1 to 12. Clearly, the cost decreases as the number of controllers decreases. If $k \geq 4$, the cost falls within the supposed acceptable range of 500.

3.3.3   Analysis of More Topologies

We now present the results of our simulations on more topologies in the Internet Topology Zoo with the goal of accurately answering the following question: how many controllers should be used in order to minimize the cost? We employed this data set because it covers a diverse range of geographic areas, network sizes, and topologies. We used the most recent version of topologies in the Zoo and all those with nodes more than 20. For all of the topologies, we set the maximum allowable latency of each network to be equal.

Figure 3.6 depicts the impact of the number of controllers on the control network cost. The x-axis is the number of controllers, which ranges from 1 to 8. The y-axis shows the corresponding cost of controllers. We found that cost optimizations on different topologies provide similar results, and larger topologies generally require more controllers to attain the same fractional reduction in cost. As expected, using more controllers reduces cost. BtNorthAmerica, for instance, which is the largest of the topologies with more than 35 nodes and 75 edges, needs at least seven controllers. However, for most of topologies, past a certain number of controllers,

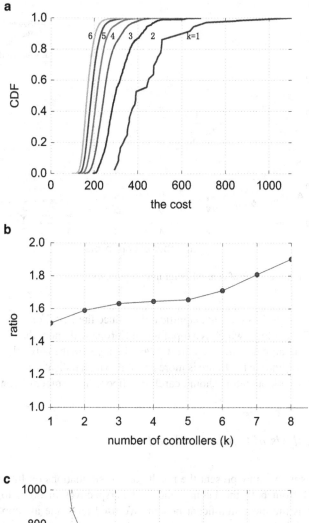

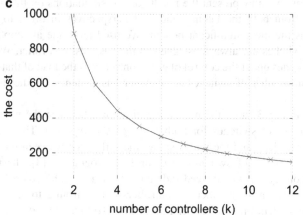

Fig. 3.5 Simulations results based on SINET4. (**a**) The CDF of the cost generated from all possible combinations of controller placements. (**b**) Ratio of randomly deploying to our proposed placement. (**c**) The cost for the number of controllers when $k = [1, 12]$

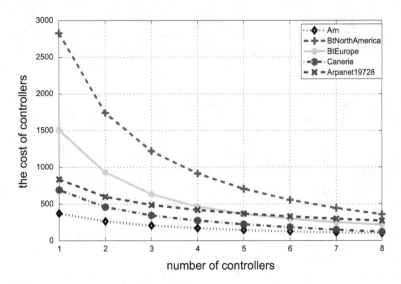

Fig. 3.6 Cost for the number of controllers for more topologies

adding more controllers does not significantly reduce the cost. For example, for the Arpanet19728 topology, which is a topology of Europe with more than 28 nodes and 31 edges, when more than three controllers are deployed in the network, the slope of the proportional lines in the figure is more gradual, which indicates a smaller benefit. Therefore, network operators should carefully choose the number of controllers.

3.3.4 Analysis of Controller Load

For comparison, we now present the results of our simulations on Internet2 OS3E and SINET4 from both the latency and cost perspectives in order to determine which one has the most significant benefit. We used the same assumption in [3]; specifically, switches are always assigned to their nearest controller. We used the number of switches under the control of each controller as the load of that controller. The maximum number of switches a controller has to control is the load on that controller.

As shown in Fig. 3.7a, b, the messaging difference between the maximum and minimum number of switches for both topologies is reported. The x-axis is the number of controllers, and the y-axis shows the difference between the maximum and minimum number of switches in control. Analogous to [3], the number of switches per controller is imbalanced and ranges from 4 to 22, based on the Internet2 OS3E topology when the number of controllers ranges from 2 to 5. Furthermore, the number of switches per controller ranges from 10 to 52, based on the SINET4 topology when the number of controllers ranges from 2 to 6.

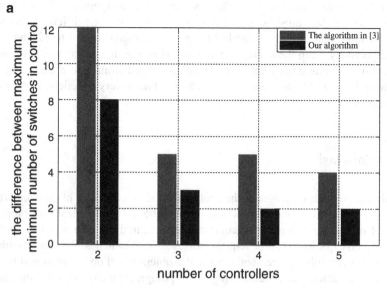

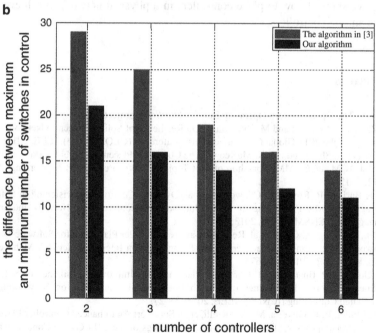

Fig. 3.7 Comparison of the assignment of switches. (**a**) The difference from both the latency and cost perspectives based on OS3E. (**b**) The difference from both the latency and cost perspectives based on SINET4

To be resilient against controller overload, the assignment of switches to different controllers must be well balanced. As expected, the difference between the maximum and minimum number of switches ranged from 4 to 12 and 14 to 29, respectively, from the latency perspective. However, the difference between the maximum and minimum number of switches ranged from 2 to 8 and 11 to 21, respectively, from the cost perspective. Clearly, this is very beneficial from a cost perspective.

3.4   Conclusion

In this chapter, we investigated the problem of SDN controller placement, which is one of the most important practical issues for SDN deployment. We proposed a novel metric for controller placement that measured the cost when controllers with limited capacity to handle request messages from switches. We first formulated the optimal controller placement as an ILP problem based on the proposed metric. We then conducted experiments using real topologies. Finally, with the simulation results, we showed how to place controllers in a physical network to achieve load balance among controllers.

References

1. ONOS: The Open Network Operating System, http://onosproject.org/.
2. Y. Zhang, N. Beheshti and M. Tatipamula. On Resilience of Split-architecture Networks. Proceedings of the 2011 Global Communications Conference (GLOBECOM). IEEE, 2011: 1–6.
3. B. Heller, R. Sherwood and N. McKeown. The Controller Placement Problem. Proceedings of the 1st ACM SIGCOMM Workshop on Hot Topics in Software Defined Networking. ACM, 2012: 7–12.
4. M. F. Bari, A. R. Roy, S. R. Chowdhury, et al. Dynamic Controller Provisioning in Software Defined Networks. Proceedings of the 9th International Conference on Network and Service Management (CNSM). IEEE, 2013: 18–25.
5. Y. Hu, W. Wang, X. Gong, et al. Reliability-aware Controller Placement for Software-Defined Networks. Proceedings of the International Symposium on Integrated Network Management (IM). IFIP/IEEE, 2013: 672–675.
6. M. Guo and P. Bhattacharya. Controller Placement for Improving Resilience of Software-Defined Networks. Proceedings of the 4th International Conference on Networking and Distributed Computing (ICNDC). IEEE, 2013: 23–27.
7. L. F. Muller, R. R. Oliveira, M. C. Luizelli, et al. Survivor: An Enhanced Controller Placement Strategy for Improving SDN Survivability. Proceedings of the 2014 Global Communications Conference (GLOBECOM). IEEE, 2014: 1909–1915.
8. S. Lange, S. Gebert, T. Zinner, et al. Heuristic Approaches to the Controller Placement Problem in Large Scale SDN Networks. IEEE Transactions on Network and Service Management, 2015, 12(1): 4–17.
9. V. Arya, N. Garg, R. Khandekar, et al. Local Search Heuristics for K-median and Facility Location Problems. SIAM Journal on Computing, 2004, 33(3):544–562.

10. K. Jain and V. V. Vazirani. Approximation Algorithms for Metric Facility Location and K-median Problems Using the Primal-dual Schema and Lagrangian Relaxation. Journal of the ACM, 2001, 48(2): 274–296.
11. Internet2 open science, scholarship and services exchange. http://www.internet2.edu/net work/ose/.
12. Science and Information Network (SINET). http://www.sinet.ad.jp/.
13. S. Knight, H. X. Nguyen, N. Falkner, et al. The Internet Topology Zoo. IEEE Journal on Selected Areas in Communications (JSAC), 2011, 29(9):1765–1775.
14. N. Handigol, B. Heller, V. Jeyakumar, et al. Reproducible Network Experiments using Container based Emulation. Proceedings of the 8th International Conference on Emerging Networking Experiments and Technologies. ACM, 2012: 253–264.
15. POX Controller. http://www.noxrepo.org/pox/about-pox/.

Chapter 4
Management System of Heterogeneous Software-Defined Networking Controllers

Abstract In the previous chapter, we studied the controller placement problem to improve the performance of the distributed control plane. When placing multiple controllers into their proper positions, the management of controllers is an important problem. In the distributed control plane, each software-defined networking (SDN) controller has its own control domain. To realize entire network control and management, these controllers must communicate with each other. However, there are no standard communication interfaces between controllers in the standard architecture of SDN. In particular, when there are heterogeneous controllers in a large-scale network, it is very difficult to coordinate these controllers to improve control plane performance. Moreover, heterogeneous controllers provide entirely different application programming interfaces (APIs) for users, leading to difficulties in management and application development. To address this problem, we propose a controller management system that consists of a heterogeneous controller management (HCM) module, domain relationships management (DRM) module, database module, and front-end module. This system can generate a global network view by collecting network information from a group of controllers while providing unified APIs for application developers that shield the differences among heterogeneous controllers.

To deploy software-defined networking (SDN) in distributed datacenters, it is necessary to build a logically centralized, physically distributed control plane. The entire network is divided into several domains. Each controller covers one domain to control switches in that domain. To realize centralized control of the entire network, an information exchange should be implemented among different domains.

In existing work, most SDN controllers do not provide the communication API. To address this issue, the Internet engineering task force (IETF) developed a protocol SDN to provide an interface for information exchange between SDN controllers [1]. Lin et al. proposed a west-east bridge for SDN inter-domain communication [2, 3]. Helebrandt et al. designed a new architecture that provides inter-domain communication based on a vendor neutral communication protocol [4]. These works focus on designing complex interfaces to achieve the goal of inter-domain communications. To reduce the complexity in interface design, a new distributed controller named the Open Network Operating System (ONOS) was

© The Author(s) 2016

H. Qi, K. Li, *Software-Defined Networking Applications in Distributed Datacenters*, SpringerBriefs in Electrical and Computer Engineering, DOI 10.1007/978-3-319-33135-5_4

built by borrowing the idea of server cluster [5]. When each domain is controlled by an ONOS controller, information exchange can be easily realized among different domains.

It is difficult, however, to implement communication between different domains when they are controlled by heterogeneous controllers. Little research has been conducted on communication between heterogeneous controllers and the management of heterogeneous controllers. To address this problem, we propose a promising controller management system based on our previous work [6]. This system consists of four sub-modules: the heterogeneous controller management (HCM) module, the domain relationships management (DRM) module, the database module, and the front-end module. We then design and implement a prototype system that can manage Floodlight, RYU, POX, and other famous open-source SDN controllers. Moreover, the system can shield the differences among these heterogeneous controllers. Users can operate heterogeneous controllers using a uniform graphical user interface (GUI).

The remainder of this chapter is organized as follows. Section 4.1 proposes the architecture of the controller management system. Section 4.2 discusses performance measurements of the prototype system and other particular cases. Finally, Sect. 4.3 concludes this chapter.

4.1   The Architecture of the Controller Management System

Figure 4.1 shows the architecture of SDN with the controller management system. This system is built on top of the control plane, which can encapsulate the whole control plane to provide a uniform interface and front-end GUI. Users can use the uniform interface to control the data plane without considering the differences among the heterogeneous controllers.

4.1.1   The Heterogeneous Controller Management Module

The function of the heterogeneous controller management (HCM) module is to manage different controllers, such as Floodlight, POX, and Maestro. The HCM module can be viewed as a middleware between the control plane and users. Users can operate heterogeneous controllers by a uniform GUI. The HCM module can convert user operations into controller APIs to communicate with the control plane.

When the controller management system is running, the HCM module collects network information from a group of controllers to generate a global-wide network view. The collected statistics include domain, controller, flow, host, link, switch, switch port, and traffic data information. HCM transmits this information to a front-end module, i.e., a GUI, which displays information to users. Users can define new flows by adding flow entries and can set flow entry parameters in the GUI. The HCM

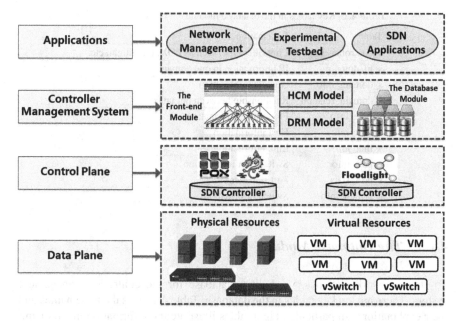

Fig. 4.1 Architecture of software-defined networking with the controller management system

module can translate this information into a Hypertext Transfer Protocol (HTTP) request based on the type of controller. The HCM then transmits the HTTP request to the control plane to install flow entries into the switches.

4.1.2 The Domain Relationship Management Module

For a distributed datacenter network, it is very important to define global inter-domain relationships. In particular, it is key for network service management, can help achieve load balancing, congestion avoidance, and fault tolerance, and prevent network configuration errors.

To address this problem, we propose the domain relationship management (DRM) module that includes two types of domain relationships: children-to-parent (c2p) and sibling-to-sibling (s2s). The c2p relationship means that each sub-domain is covered by a secondary controller while a primary controller connects to these secondary controllers. The primary controller maintains the whole network by collecting information from secondary controllers. The s2s relationship means that each sub-domain is covered by a controller while these controllers are intercon-nected. Any two controllers can directly exchange messages to maintain the whole network. In the DRM module, if multiple controllers are heterogeneous, the c2p relationship is adopted, and the s2s relationship is adopted, otherwise.

Table 4.1 Key tables in the database module

Table	Explanation
s_controller_info	Controller type, such as POX, Floodlight, etc.
s_ip_domain	IP information in one domain
s_domain_info	Domain information
s_domain_relation	Relationship between domains
s_flow_info	Flow entry information
s_host_info	Host information
s_link_info	Link information
s_port_info	Switch port information
s_port_stats	Port statistics
s_switch_info	Switch information

4.1.3 The Database Module

To store network statistics, we built a database for the controller management system. There are ten key tables in the database. Table 4.1 gives the table names and their explanations. In particular, eight tables illustrate the information in a domain: basic information, controller, flow, host, link, switch, switch port, and switch port statistics. The other two tables give information between domains. The s_ip_domain table displays the IP information in one domain. The s_domain_relation table shows the relationship between domains as c2p or s2s.

4.1.4 The Front-End Module

The front-end module provides a uniform GUI to users where they can manage the whole network more conveniently without considering complex controllers' APIs. As shown in Fig. 4.2, this module displays network topologies to users. Moreover, users can add and delete flow entries using GUI operations. For example, Fig. 4.3 shows the interface of adding flow entries. Flex and ActionScript are used to implement the GUI in this module. RemoteObject is used to implement the data interaction between this module and the server-end including HCM and DRM modules.

Although there are some existing controller systems with GUIs, there is no GUI that can show a large-scale network controlled by multiple controllers. Table 4.2 shows the comparison between the GUI of our system and those of other existing controller systems.

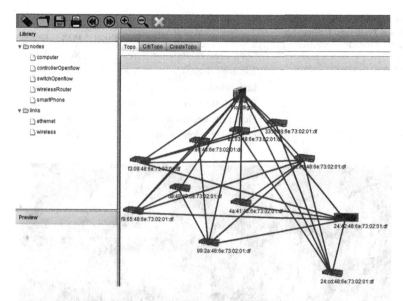

Fig. 4.2 Network topologies presentation in the controller management system

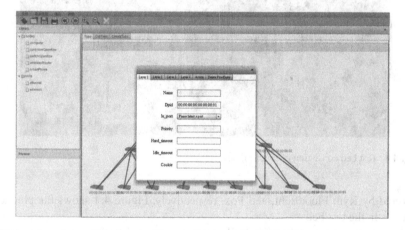

Fig. 4.3 Interface of adding flow entries in the controller management system

4.2 System Evaluation

To evaluate the proposed system, we built a fat-tree structure-based datacenter test bed including eight practical nodes. In the test bed, we used four Pica8 P-3297 switches [7], a NetFPGA card, an ONetSwitch20 card [8], and some OpenvSwitches on Ubuntu 12.04 as the forwarding devices in the data plane. To simulate multiple SDN domains, we divided the datacenter network into three domains, which were

Table 4.2 Comparison of the proposed system and other existing controller systems

	GUI		
Function	Floodlight	POX	Our system
Component information	Yes	No	Yes
Topology display	Yes	Yes	Yes
Link information	No	No	Yes
Port information	No	No	Yes
Flow table operation	No	No	Yes
Flow table information	Yes	Yes	Yes

Fig. 4.4 Test bed consisting of practical devices

covered by Ryu, Floodlight, and Pox, respectively. Figure 4.4 shows the practical devices in the test bed.

In our experiments, three controllers were connected to the controller management system. We used the apache benchmark to measure system performance. First, we compared the request completion time of the proposed system to that of the Floodlight controller. We built an application to send 100,000 flow table entry update requests and then monitored the time of request completion. From the results shown in Fig. 4.5, we can see that the proposed system finishes more than 60 % of the requests in 5 ms and more than 95 % of the requests in 10 ms. The request completion time of the proposed system is similar to that of the Floodlight controller.

Next, we measured the throughput of the proposed system with different numbers of threads. In this experiment, we compared the proposed system, Floodlight controller, and Maestro controller. In the data plane, there were 16 openflow

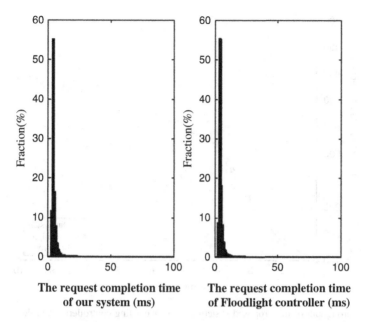

Fig. 4.5 Comparison of the request completion time between the proposed system and the Floodlight controller

switches. Figure 4.6 shows the throughput of the proposed system and other existing controllers for varying numbers of threads. We can see that the Floodlight and Maestro controllers both process less than 0.8 million requests per second with ten threads. With the same number of threads, the proposed system can process more than 0.9 million requests per second. Thus, the throughput of the proposed system is better than that of existing controllers.

Finally, we measured the latency of the proposed system. In this experiment, we analyzed the responding time with varying numbers of switches. The comparison results are shown in Fig. 4.7. According to Fig. 4.7, we can see that the proposed system has a longer responding time when connecting to less switches. The reason for this is that the proposed system interacts with the control plane. When connecting to more switches, the processing capacities of the Floodlight and Maestro controllers are limited, leading to more higher latency. Because the flexibility of the proposed system is better, it has less latency.

4.3 Conclusion

In this chapter, we designed and implemented a controller management system, which provides a uniform GUI to manage the distributed datacenter network covered by multiple heterogeneous controllers. The proposed system shields the differences among heterogeneous controllers to achieve more convenient network management.

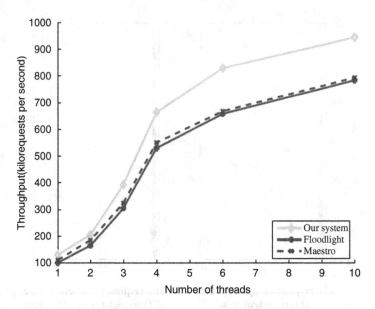

Fig. 4.6 Throughput of the proposed system and other existing controllers when the number of threads is varied

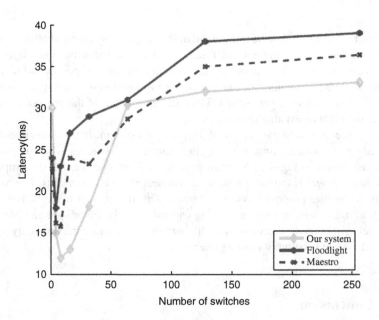

Fig. 4.7 Latency of the proposed system and other existing controllers when varying the number of switches

We also presented the architecture of the proposed system and built a practical datacenter as a test bed to evaluate the proposed system. Our experimental results show that the proposed system achieves high performance.

References

1. H. Yin, H. Xie, T. Tsou, et al. SDNi: A Message Exchange Protocol for Software Defined Networks (SDNS) across Multiple Domains. [Online]. Available: https://tools.ietf.org/html/draft-yin-sdn-sdni-00.
2. P. Lin, J. Bi, Z. Chen, et al. WE-bridge: West-East Bridge for SDN Inter-domain Network Peering. Proceedings of the 2014 IEEE Conference on Computer Communications Workshops (INFOCOM Workshops). IEEE, 2014: 111–112.
3. P. Lin, J. Bi, S. Wolff, et al. A West-East Bridge based SDN Inter-domain Testbed. IEEE Communications Magazine, 2015, 53(2): 190–197.
4. P. Helebrandt and I. Kotuliak. Novel SDN Multi-domain Architecture. Proceedings of the 12th International Conference on Emerging eLearning Technologies and Applications (ICETA). IEEE, 2014: 139–143.
5. S. S. Hayward, S. Natarajan and S. Sezer. A Survey of Security in Software Defined Networks. IEEE Communications Surveys & Tutorials, 2015, DOI: 10.1109/COMST.2015.2453114.
6. H. Yu, K. Li, H. Qi, W. Li and X. Tao. Zebra: An East-West Control Framework For SDN Controllers. Proceedings of the 44th International Conference on Parallel Processing (ICPP). IEEE, 2015: 610–618.
7. The Pica8 Openflow Switch, http://www.pica8.org/.
8. The ONetSwitch20 from Meshsr company. http://www.meshsr.com/.

Chapter 5
Conclusions and Future Research Topics

Abstract In this chapter, we summarize the studies presented in this book. Then, we discuss some future research topics related to software-defined networking (SDN).

5.1 Conclusions

In this book, we discussed the design and deployment of SDN applications in distributed datacenters. We focused on the SDN-based request allocation mechanism, SDN controller placement strategy, and SDN controller management system in distributed datacenters. The specific contributions are as follows:

- We found that the central control provided by SDN is an effective way to address the request allocation problem in distributed datacenters. In light of this fact, we proposed a joint optimization model for request allocation, which considered both service providers and end-users. To give the solution, we proposed a Nash bargaining solution (NBS)-based algorithm. According to the simulation experiments with real-world traces, we verified the effectiveness of the proposed algorithm. Based on this algorithm, we designed the SDN-based request allocation mechanism.
- To maximize the benefits of SDN, we must effectively deploy SDN in distributed datacenters. To address the problem of SDN controller placement in SDN deployment, we proposed a novel placement metric that considered the cost of controllers when their processing capacities were limited. We presented an integer linear program (ILP)-based optimization model. To solve this problem, we designed an effective approximation algorithm as an SDN controller placement strategy. According to the simulation experiments, which were based on many real topologies, we demonstrated the high performance of the proposed SDN controller placement strategy.
- In large-scale distributed datacenter networks, it is necessary to deploy multiple SDN controllers while maintaining a global network view from these distributed controllers. To achieve this goal, we designed and implemented a management system of heterogeneous SDN controllers. This system shields differences among the heterogeneous controllers to provide a uniform graphical user interface (GUI)

© The Author(s) 2016

H. Qi, K. Li, *Software-Defined Networking Applications
in Distributed Datacenters*, SpringerBriefs in Electrical and Computer Engineering,
DOI 10.1007/978-3-319-33135-5_5

for users. To evaluate the proposed system, we built a test bed consisting of practical openflow switches and servers. The experimental results verified the high performance of the proposed SDN controller management system.

5.2 Future Research Topics

In the next few years, SDN will perpetually be a hot point in future network areas; thus, we plan to make further progress toward solving existing SDN problems. It is our opinion that the important future research topics related to SDN include the following three areas:

- In the data plane, OpenFlow is not the only option. It is difficult to design a better southbound application programming interface (API) to improve the flexibility of the data plane. Moreover, the data plane should have minimal computational capabilities to reduce the burden of the control plane.
- In the control plane, the deployment of SDN in large-scale networks remains a big problem, especially for improving the expansibility of the control plane. When realizing global control of large-scale networks, a simple and effective global optimization algorithm is necessary.
- In the application plane, there is lack of high-level network programming language for building network innovation applications. A unified and powerful northbound API is also needed. It is difficult to realize application driven network control to take full advantage of SDN.

Printed in the United States
By Bookmasters